NASA

THE COMPLETE ILLUSTRATED HISTORY

To my favorite twins:
Christine M. Gorn, my wife, and
Leo Joseph DeKhors, my brother-in-law

FOREWORD BY **BUZZ ALDRIN**

NASA

THE COMPLETE ILLUSTRATED HISTORY   **MICHAEL GORN**

MERRELL
LONDON · NEW YORK

Contents

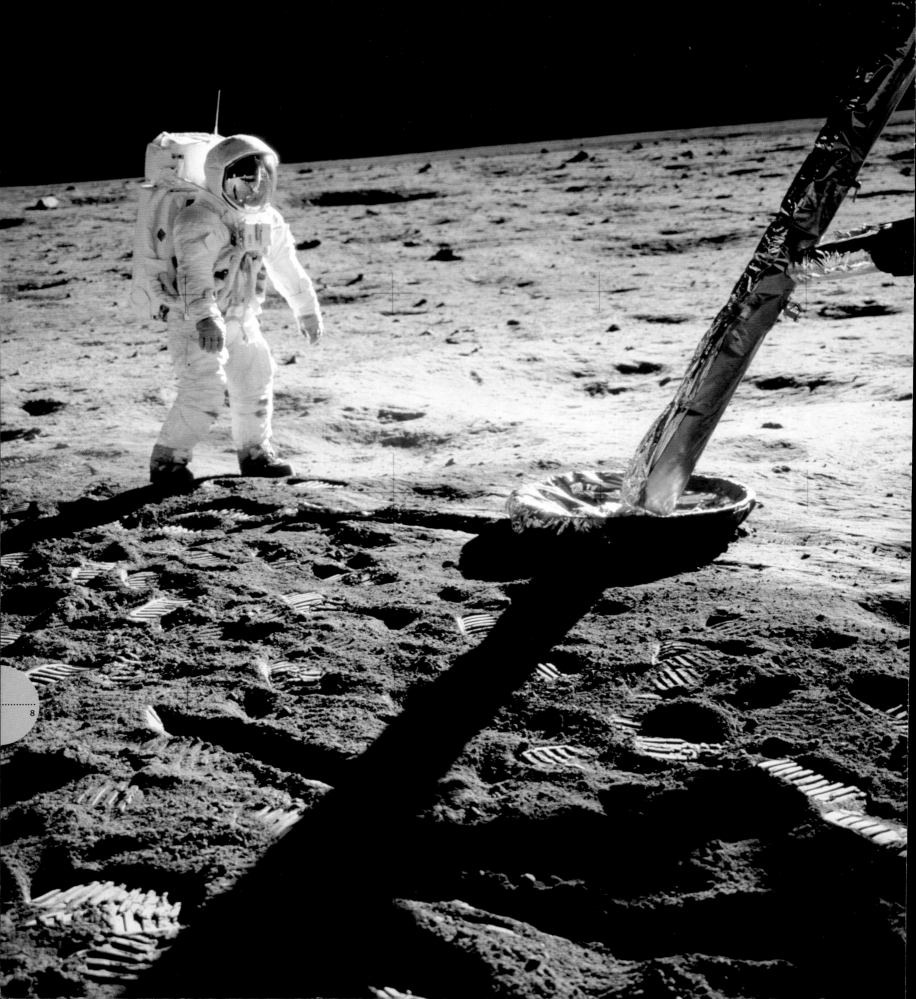

To me, the story of NASA is essentially a story of inspiration. Of course, the agency has also achieved many practical things of great value. To name only a few, the agency's scientists raised the first red flags about global warming. Its earth sensing satellites revealed facts about conditions on Earth—such as weather, topography, and forestation—not known before. NASA engineers have worked diligently to limit the noise of jetliners and to minimize the effects of sonic booms. The list of the agency's technological accomplishments having direct application to the human race is a very long one.

In fact, NASA really *began* for practical reasons. It was conceived during the Cold War in order to win the space race with the U.S.S.R., one of many desperate skirmishes between the two superpowers. In the process, President Kennedy gambled everything on the Apollo program and deserves the gratitude of the nation for giving Americans a visionary purpose in space. But this benefit occurred almost incidentally. The president actually launched Apollo for the practical objectives of proving America's technical superiority over the Soviets and showing the world that the tide of history rose with the West, rather than with Communist ideology. The Moon only represented the finish line on one front of this global struggle.

Still, in the pursuit of these ideological goals, NASA gave the world inspiration, and I was privileged to be a part of it. During my 290 hours of spaceflight (eight of them in extravehicular activity), I experienced this sense of exhilaration and wonder many times. Looking back at Earth as Neil Armstrong and I stood on the Moon, I will never forget our planet as a brilliant jewel, set against a black velvet sky. I am always impressed by something

people of every walk of life have told me repeatedly since that flight in 1969. They almost never ask me about what rocks we picked up, or what Neil, Mike Collins, and I said to each other. Instead, they want me to relive with them where they were or what they were doing when Neil and I walked out of the *Eagle* as Mike flew overhead. And when you multiply that times millions of people, that's why humanity does things.

NASA: The Complete Illustrated History recounts the history of the American space agency in a way that I hope will inspire you, too. Its beautiful pictures and crisp text remind me of the many great things the United States—and its international partners—have been able to achieve. But the book also reminds me that spaceflight is only about fifty years old and still in its infancy. I think it is important to remember that the future of space exploration is entirely in human hands. Do we content ourselves with space as a place of practical opportunities such as telecommunications breakthroughs, advanced weather prediction, and global mapping? Or do we reach for inspiration? Personally, I would like as many human beings as possible to experience—even indirectly—what I did in July 1969. That is why I am hopeful that the spacefaring nations of the world (growing larger in number almost with each passing year) will set their sights on Mars, our brother planet. This adventure will not only stretch our technological capacities but will represent one of the great events of the twenty-first century. In my judgment, the survival of humans on the surface of Mars is one of the noblest things earthlings could ever witness.

Buzz Aldrin
Los Angeles, California, U.S.A.

Astronaut Buzz Aldrin, lunar module pilot, walks on the Moon near a leg of the lunar module during Apollo 11 *extravehicular activity. Astronaut Neil Armstrong,* Apollo 11 *commander, took this photograph with a 70 millimeter lunar surface camera. The astronauts' footprints are clearly visible in the foreground.*

Foreword by Buzz Aldrin

1

Before NASA
The National Advisory Committee
for Aeronautics

↑ *An aerial view taken in 1950 of the Langley Research Center, Hampton, Virginia, successor of the Langley Memorial Aeronautical Laboratory. Along the shore of the Back River, at the bottom of the photograph, is the original layout (the East Area). The top of the picture illustrates that part of Langley constructed early in World War II (the West Area).*

On a late spring day in 1920, a select group of Washington insiders, local dignitaries, and air enthusiasts gathered for the dedication of America's first national aeronautical laboratory. Hewn from the hard clay, the broad marshes, and the dense forests native to the countryside outside of Hampton, Virginia, the Langley Memorial Aeronautical Laboratory received full honors on its opening. Viewers watched in fascination as Brigadier General William "Billy" Mitchell, the Army Air Service's Director of Military Aeronautics, led an aerial extravaganza, culminating in the over flight of a mighty formation of twenty-five combat aircraft. Rear Admiral David Taylor, the Chief Constructor of the Navy, joined other prominent uniformed and civilian officials who praised the new facility. Taylor called it no less than "the shrine to which all visiting aeronautical

engineers and scientists will be drawn." The guests then saw for themselves this still unfinished laboratory, named for the famed aeronautical pioneer (and Secretary of the Smithsonian Institution) Samuel P. Langley.

In actuality, the crowd did not have much to see on June 11, 1920. Few research buildings had been completed, but the tour did feature one impressive piece of machinery—the 5 Foot open-end Atmospheric Wind Tunnel, patterned after that of the British National Physical Laboratory. Although small and not too reliable, it sprang to life at the command of its technicians, impressing those present with the roar of the engine, the pulse of the rotating blades, and the force of the onrushing air. Yet, if the visitors walked by the new hangar and flight line, their enthusiasm may have dampened.

↑ *The first Langley wind tunnel, viewed by dignitaries during opening ceremonies of the laboratory in June 1920. Inefficient due to its open-end design but costing only $38,000, the 5 Foot Atmospheric Wind Tunnel proved to be of limited value as a research facility.*

Here stood the laboratory's entire air armada: three war surplus JN4H Jenny aircraft borrowed from the Army Air Service.

Modest as the reality may have been, this day had been a long time coming and, in fact, might never have occurred at all. It constituted something of a miracle, and even more so in light of Langley's future eminence. Indeed, nothing at its humble start predicted the future importance either of the laboratory or of its parent organization, the National Advisory Committee for Aeronautics (NACA). Eventually, the NACA assumed an honored role as the foremost federal entity devoted exclusively to aeronautical research and development. Many years later, the NACA became the essential ingredient among the sundry bureaucratic components

that comprised the even more prominent National Aeronautics and Space Administration (NASA).

Yet, despite its impact on the history of flight, it originated as an afterthought. The law establishing the NACA almost went unnoticed, a two paragraph rider to the Naval Appropriations Act of 1915. The result of a campaign of air-minded citizens rather than the will of the White House, the Congress, the civil service, or industry, the NACA relied on no lobbyists to bring it to fruition. As a consequence, it not only began without a mandate but also set out on a starvation budget of only $5000 per annum for its first five years. Thus, in contrast to contemporary European laboratories, which represented true national commitments to aeronautical research, the United States launched a makeshift organization the

↑ About to take wing from the maiden Langley flight line in spring 1919, these two Army Air Service JN4H Jenny aircraft comprised two-thirds of the NACA's air fleet during its first years. Even as the laboratory underwent construction, these vehicles conducted important flight research in the skies above Hampton, Virginia.

←···· Cutting-edge flight as it existed only thirteen years before Congress authorized the National Advisory Committee for Aeronautics (NACA). Wilbur and Orville Wright conducted flight research on their glider design in October 1902 at Kill Devil Hills, North Carolina.

longevity and influence of which seemed doubtful at best.

Just twelve years before the NACA came into being, on the morning of December 17, 1903, Orville Wright flew at Kitty Hawk, North Carolina, on the wings of a frail airplane for twelve seconds, covering a distance of 120 feet (36.5m). The machine, flying above the sands and dunes, finally shuddered to a halt on the beach. Despite the epochal meaning of this achievement, the world remained largely ignorant of it until 1908. In part, this five year delay occurred because the Wrights hid their secret as they sought to patent the technology. But just as important, the very improbability of the event defied belief, causing those who heard about it—living before the time of paparazzi and instant digital images—to dismiss it as an impossibility. Only when the Wrights finally flew for large audiences did the reality of their breakthrough become inescapable.

Consequently, the NACA actually received its birth certificate a mere seven years after the first powered flight became acknowledged universally. Indeed, very soon after this realization, many of the industrialized powers comprehended the potential of aeronautics both for good and for ill. World War I confirmed the suspicions of the more pessimistic. Initially fielded for reconnaissance purposes, aircraft soon became significant offensive weapons. Above the trenches of the Western Front, Sopwith Camels, Fokkers, and Nieuports contested for the skies over the battlefields. Meanwhile, bombers—like Igor Sikorsky's famed multi-engine giant, *Il'ya Muromets*—threatened terror and destruction from above.

Although the Wrights' contribution should have given their countrymen a long lead in the pursuit of flight, it was the European countries—more accustomed than the Americans to regarding the state as an ally of modernization—that forged powerful national institutions dedicated to aeronautical research. France acted first, unveiling the Central Establishment for Military Aeronautics at Chalais-Meudon, near Paris. In Russia, the Aerodynamic Institution in Kuchino opened and aligned itself with the University of Moscow. In Germany, the Aerodynamical Laboratory came into existence under the auspices of Gottingen University. Finally, the United Kingdom took perhaps the boldest step, initiating at significant public cost the British Royal Aircraft Factory at Farnborough. Furthermore, in France, Germany, and Britain these aeronautical institutions received the advice and backing of distinguished advisory boards comprised of eminent civil servants, university professors, and industrialists who counseled their governments about technical developments at home and abroad.

The formation of the NACA, and the subsequent construction of a laboratory to carry out its research, happened more haltingly than was the case with its European counterparts. Just a few weeks after the 1915 Naval Appropriations Act brought the NACA into being, the NACA's Main Committee (appointed by President

←···· The founding members of the National Advisory Committee for Aeronautics met in April 1915 in the offices of the Secretary of War. Elected temporary chairman, Brigadier General George P. Scriven (seated, third from left), Chief of the U.S. Army Signal Corps, played a fundamental role in locating the Langley laboratory in Hampton, Virginia.

Woodrow Wilson) convened for the first time in Washington, D.C., under the leadership of Brigadier General George P. Scriven, Chief of the U.S. Army Signal Corps (the guiding hand of the service's aviation activities). Like the European advisory committees, it too consisted of distinguished but unpaid members, in this instance enlisted from the Smithsonian Institution, the armed forces, the National Bureau of Standards, the Weather Bureau, and the private sector. The Main Committee members (at first no more than twelve, later expanded to fifteen) selected from among their number a smaller body they called the Executive Committee and invested in it the authority to meet often and to run the practical affairs of the organization. These individuals, in turn, recruited scientists and engineers to serve on a series of NACA technical panels devoted to such fields as propulsion, aerodynamics, structures, and so on.

Of course, none of these activities mattered without a laboratory to conduct research, indispensable to fulfilling the recommendations of the Executive Committee's technical panels. So, in 1916 the members of the Main Committee requested funds from Congress to construct a facility. More importantly, the influential Secretary of the Smithsonian Institution, Charles Walcott—who also served as the first chairman of the NACA's Executive Committee—appealed to Congress personally. Walcott asked for a budget of $85,000 in fiscal year 1917, $53,000 of which was to initiate a national aeronautics laboratory. Under the stress of war, Congress approved the request.

Meanwhile, General Scriven took on the challenge. He received orders from his Army superiors to designate a group of promising locations on which to construct an experimental station for military aeronautics. After assessing fifteen sites, the selection board chose

1650 acres (668 ha) just north of Hampton, Virginia, on an Army post later named for Samuel P. Langley, former Secretary of the Smithsonian Institution and aeronautical researcher thwarted by the Wrights to be first aloft. Scriven, well aware of the NACA's pressing need to find a home, invited its Main Committee to scrutinize the parcel and designate a portion for the NACA laboratory. The area looked promising. It enjoyed relatively favorable weather, close proximity to the seat of government in Washington, D.C., and plenty of skilled labor down the road at the naval shipyard in Newport News. Without other viable options, the NACA leaders elected to join the venture.

Many came to regret the decision, at least in the short term. Between summer 1917 and the opening of the Langley Memorial Aeronautical Laboratory three years later, officials involved with the construction encountered nothing but obstacles and hardships. First, the individuals hired to clear the land and build the structures met any number of miseries. "The muddiest mud, the weediest weeds, the dustiest dust, and the most ferocious mosquitoes" bedeviled the work gangs.[1] They cut down endless trees and blasted endless tree stumps. The heat and humidity of the region took their toll. The marshy landscape seemed to absorb every shovelful of dirt turned against it. As a consequence, actual building proceeded slowly; the first structures began to appear only at the end of 1917.

Meanwhile, the Army became less supportive. Constrained by the operational pressures of World War I, it abandoned its earlier commitment to base its aerial experimental research at Langley, instead transferring the projects to McCook Field in Dayton, Ohio. This decision not only left the NACA without an anticipated research partner but also meant that, because Langley now

One of the two original side-by-side Langley hangars, under construction in 1922. Eventually, during the 1920s, this hangar and the one adjacent to it would shelter a growing laboratory fleet, that included such varied workhorses as a DeHavilland 4, a Fokker D-VII, a Sperry M-1 Messenger, and a Vought VE-7.

An aerial perspective on the Langley administration building, photographed in 1924. Even four years after the laboratory opened, visitors and employees still had to cope with a rough, unfinished landscape, exemplified by the mud surrounding this structure.

became a training base, its officers could evade the NACA's repeated requests for formal assignment of land for the laboratory, further hampering the construction timetable. The last straw came in summer 1919 when Army authorities, pressed by the postwar constriction of facilities, cut off NACA access to military housing and utilities.

As a consequence, frustration and disappointment intensified during this period and some NACA officials even lobbied for the Langley project to be abandoned in favor of building a laboratory on Bolling Field in Washington, D.C. Congress declined to entertain this plan after sinking so much into Hampton. The work continued, and in June 1920 the Langley Laboratory held an impressive opening ceremony complete, paradoxically, with full Army pageantry.

If the bricks-and-mortar part of the NACA's early existence raised much anxiety, its founding figures gave much assurance. Indeed, the NACA's fortunes improved notably when it found two very able, but very different leaders. As luck had it, both stayed in their posts for a generation or more, long enough to protect, nurture, and finally expand the NACA.

The participation of Professor Joseph Sweetman Ames of Johns Hopkins University represented a precious asset to the fledgling NACA. Ames received an appointment in 1915 from President Woodrow Wilson to sit as one of the original members of the NACA Main Committee. Energetic and forthright, he assumed a role of leadership almost immediately. The NACA recognized his value in 1919 when it made him chair of the Executive Committee after Charles Walcott moved on to lead the Main Committee. During the eighteen years he held this post, Ames became the NACA's indispensable man, defining the research agenda of the young agency.

Meantime, after a succession of administrative roles on

Joseph S. Ames

Joseph S. Ames, one of the nation's most influential physicists and college administrators, had a profound impact on the NACA. Born in Vermont in 1864, the son of a physician, he grew up in Minnesota. Although Latin and Greek appealed to him most, in 1883 he enrolled as an undergraduate physics major in a new university endowed by Baltimore's wealthy Quaker merchant, Johns Hopkins. After a year of graduate study in Berlin, he received the doctorate in physics from Hopkins in 1890. Ames began a contented home life in 1899 when he married Mary Harrison, an attractive, graceful Baltimore widow with three children. Her friendly manner acted as counterpoint to his assertive personality (attributable, perhaps, to a stammer with which he struggled all his life). Neither a theorist nor an experimentalist, Ames instead became one of the most important synthesizers in American physics. In keeping with his encyclopedic knowledge, he became famous for his twelve-volume edition of *Harper's Scientific Memoirs* and co-authored a popular book about laboratory practice entitled *A Manual of Experiments in Physics*. Upon graduation from Johns Hopkins University, Ames won a teaching position there and rose rapidly, becoming professor of physics in 1898 and director of the university's Physical Laboratory in 1901. Hopkins further recognized his administrative talents by naming him acting president between 1912 and 1914, and president in 1929. Meanwhile, Ames continued to teach and lecture, attracting superior students such as Hugh L. Dryden, the second and last director of the NACA. Joseph Ames died in 1943, and the Ames Research Center in Sunnyvale, California, is named in his honor.

←··· *The NACA Main Committee meeting at the agency's new headquarters in Washington, D.C., in May 1920, the month before the Langley laboratory began operations. Among the important figures seated around the table is Joseph S. Ames (second from the right), who played an indispensable role in guiding the research agenda of the fledgling NACA. Two other notables are Orville Wright (fourth from the left) and Secretary of the Smithsonian Institution Charles D. Walcott (far right).*

George W. Lewis

The son of a sales representative for the American LaFrance Fire Engine Company, George W. Lewis (born in 1882) from Ithaca, New York, earned bachelor's and master's degrees in mechanical engineering from Cornell University. An opportunity to join the NACA's technical panel on power plants occurred in 1918, probably through the influence of William Frederick Durand. It was here that Joseph Ames met Lewis. No mean leader himself, Ames recognized a certain charisma in the younger man. Stocky, relaxed, and a charmer who demonstrated modesty, affability, and determination in equal parts, Lewis seemed the ideal choice to be the NACA's Director of Research. With Ames's backing, the NACA Executive Committee hired the thirty-six year old at a salary of $5000 per year. Lewis reported to work in November 1919 and remained in the same position until 1947, when he resigned due to ill-health. Ames's and Durand's hunch about Lewis's suitability proved entirely right. In short order, he transformed himself into the classic Washington insider. As comfortable in the halls of Congress as in the corridors of the federal bureaucracy, Lewis understood the importance of winning political allies for the NACA, but at the same time succeeded in building and structuring the resources of Langley in order to satisfy the objectives of Ames and his committees in Washington. Together, Lewis and Ames—two strikingly contrasting individuals—guided the fortunes of the NACA so that the young engineers at Hampton, insulated from the entanglements of politics, might pursue their research in an atmosphere of freedom. George W. Lewis died in 1948, after which the NACA's Cleveland, Ohio, Aircraft Engine Research Laboratory became known as the George W. Lewis Flight Propulsion Laboratory.

----> *George W. Lewis led the NACA as Research Director from 1919 to 1947. Lewis did more to inform the essential character of the NACA than any other person. Its focus on engineering solutions, rather than abstract theory; its atmosphere of free inquiry; and its low public profile all stemmed from Lewis's leadership.*

campus, Ames assumed the presidency of Johns Hopkins University in 1929. Two years earlier, the sixty-three year old scientist became chairman of the NACA Main Committee. During the succeeding years, Ames embodied the NACA. Indeed, from 1927 to 1937 he simultaneously led both the Executive and the Main Committees, making an indelible imprint on the institution by advising on the selection of technical personnel for the committees and the laboratory, and by steering Langley's research agenda. Eventually, under the dual load of Hopkins and NACA responsibilities, Ames's health collapsed. He suffered a paralytic stroke in 1936, resigned from the Main Committee in 1939, and—after twenty-four years' continuous service *without salary*—died in June 1943.

If Joseph Ames symbolized the NACA's intellect, the other founding figure represented its spirit. Members of the Executive and Main Committees recognized from the start that the organization required a full-time technical director to further its interests in Washington, D.C., while at the same time supervising the Langley laboratory. Committee members tried without success to interest several prominent prospects. As morale sank during the dismal situation in 1919, the need to find a good recruit became acute. Stanford University's William Frederick Durand (who acted briefly as Main Committee chair between Scriven's and Walcott's tenures) found a highly promising candidate. Durand met him during World War I at a Philadelphia aeronautical foundation called Clarke Thomson Research, where he worked as chief engineer in the fields of supercharging and gas turbine engines. After a brief period of scrutiny, the engaging and worldly George W. Lewis became the NACA's first Director of Research.

Once in office, Lewis's most pressing task involved the appointment of top quality engineering and scientific talent. He started with his own staff in Washington, by 1923 choosing seventeen individuals to make up the NACA headquarters. But overall, during his first half dozen years Lewis encountered anything but sunny skies in his personnel choices. To begin with, Langley needed leadership. Lewis discovered quickly that he could not manage the facility from Washington, even if he wanted to. At first, he solved the problem by dividing the laboratory into three categories, each with its own manager: a senior engineer who reported to the Power Plant Committee; a chief physicist responsible to the Aerodynamic Committee; and a chief clerk, guided by the Personnel Committee. As late as 1923, no one in Hampton oversaw all three parts. This structure disappeared when Lewis asked Leigh Griffith, an able Californian in charge of the Power Plant Committee, to be Langley's first Engineer-in-Charge; that is, the laboratory's initial director. But Griffith soon ran afoul of the NACA's first paid employee,

[Top left] *Two more figures of key importance in the annals of the early NACA: Leigh Griffith, Langley's first Engineer-in-Charge (far right) and John Victory, the NACA's Executive Secretary (far left). After a bureaucratic flare-up with the fussy Victory, Griffith left the NACA, to be succeeded as Engineer-in-Charge by the more phlegmatic Henry J.E. Reid.*

[Top right] *Henry J.E. Reid, Langley Engineer-in-Charge, working in his office in 1928. A respected instrumentation engineer in his own right, Reid found it easy to make friends and avoided making enemies, valuable attributes in a laboratory teeming with energetic young researchers eager to advance themselves.*

[Middle] *The first employee to be hired by the National Advisory Committee for Aeronautics, John Victory served at headquarters as Executive Secretary under George W. Lewis and then under Hugh L. Dryden. Victory liked to wield power, which grew as he perfected the NACA bureaucracy during Lewis's tenure. His influence waned after Dryden's succession in 1947.*

[Bottom] *One of the few professional women employed by the early NACA, Pearl I. Young arrived at Langley in 1922. At that time there was only one other female physicist working in the entire federal government. After seven years in the lab's Research Division, Young received instructions from Engineer-in-Charge Henry Reid to form a Technical Publications Office and to serve as its chief.*

←⋯↑ Shown here, two views of the Atmospheric Wind Tunnel, completed for the opening of the Langley Memorial Aeronautical Laboratory in 1920. In the exterior picture, two technicians stand at the intake end of the tunnel, a 5 foot (1.5 m) diameter, open-ended type. Inside is the tunnel control room, in which a model of a JN4H Curtiss Jenny is mounted for testing.

John Victory, the organization's prickly and high-handed Executive Secretary. The two men tangled over simple clerical procedures, and Griffith returned to California permanently in 1925.

Lewis decided to appoint the new Engineer-in-Charge from within Langley's ranks, a choice that almost guaranteed a youthful successor to the forty-four year old Griffith, given the preponderance of engineers in their twenties and thirties at Langley. Among them, Henry J.E. Reid, the chief of instrumentation research and development, caught Lewis's eye. Reid first came to Langley the year after the lab opened, having earned a bachelor's degree in electrical engineering from Worcester Polytechnic Institute. When Lewis chose the twenty-nine year old, he made a wiser choice than he might have known. First, like Ames and Lewis, Reid was to commit the bulk of his life to the new organization, retiring from NASA in 1960. But even more important than longevity, Reid embodied the well-rounded researcher who also possessed an instinct for effective management. Throughout his years of administrative service, he retained his interest in instrumentation and continued to

contribute to the field. At the same time, he got along well with Ames, Lewis, and John Victory—as well as the assorted temperaments who worked for him—due mostly to his judicious style and his ability to navigate among the strong and sometimes eccentric personalities at Langley. At the same time, he ran the laboratory efficiently, and according to bureaucratic norms. But perhaps most importantly, Reid understood that his ambitious, energetic, and youthful researchers required the liberty to conduct their investigations with as little interference, and as much autonomy, as possible.

No one tried the patience of the phlegmatic Reid like Dr. Max Michael Munk, one of the NACA's most able yet most temperamental minds. Indeed, Munk's highly personal style and acerbic tongue posed a challenge to the evolving NACA ethos. Rather than permit the organization to become the preserve of egos and prima donnas, Ames, Lewis, and Reid structured the agency so that it functioned regardless of the individuals involved. As Executive and Main Committee chair, Joseph Ames shaped the recommendations of the NACA technical committees in Washington, D.C., transmitting them to

BUILT BY THE
N.N.S.& D.D.C.
NEWPORT NEWS.
VA.

WEIGHT. 166760. LBS.

N   N   S   &   D   D   CO   41

Max Munk's Variable Density Wind Tunnel, seen here in side and end perspectives upon its arrival at Langley from the nearby Newport News Naval Shipyard in February 1922. Its formidable appearance and size—dwarfing the two technicians standing next to it—gave away its function: a big tank capable of withstanding high pressures.

George Lewis as Research Authorizations. Lewis, in turn, sent the Research Authorizations to Henry Reid with some suggestions about implementation, after which the Engineer-in-Charge assigned the projects to his staff.

The German-born Munk may have chafed at the good order of the NACA bureaucracy, but the magnitude of his scientific contributions certainly cannot be denied. His most influential technical insight involved the scaling effects encountered in wind tunnel models. It seemed that when engineers used small models to conduct tunnel research, the data did not necessarily coincide with the aerodynamic characteristics of full-sized aircraft of the same design. The NACA already had experienced such a problem. The Atmospheric Wind Tunnel (also called NACA Wind Tunnel Number 1), completed in 1920, had been modeled closely on the one located at the British National Physical Laboratory. Just 5 feet (1.5m) in diameter, it accommodated models no bigger than 3½ feet (1m) and produced data that differed widely from that derived from full-scale, flying aircraft. Moreover, the NACA tunnel followed the open circulation design of its British counterpart, an outmoded system supplanted by the more efficient closed circuit popularized in Germany.

Munk proposed to test model aircraft in a new way. By building a sealed tunnel to hold air that was pressurized, he felt certain that the data would rival the accuracy of

that acquired from full-scale aircraft flying in the atmosphere. The NACA Executive Committee approved the construction of such a machine in 1921, and a year later Langley employees watched as a formidable looking tank—the shell of the Variable Density Tunnel—arrived by rail from the Newport News Shipbuilding and Dry Dock Company. Munk began to appear occasionally at Hampton to oversee the fabrication, which was completed in October 1922. The initial experimental airfoil research conducted on the Variable Density Tunnel in 1923 confirmed Munk's intuition, and news spread of a wind tunnel that employed small models yet achieved data that was almost as accurate as real aircraft. As a consequence, Munk established himself and the NACA in the front ranks of international aeronautical research.

Unfortunately, Munk made enemies as easily as George Lewis made friends, and the German's technical achievements carried a high human cost. Munk stirred hard feelings during each of his sporadic visits to Hampton. In part, his German accent and stubborn insistence on European research techniques angered some with fresh memories of World War I. Also, his haughtiness and inability to accommodate other points of view alienated his co-workers, as did his habit of treating them like intellectual inferiors. Clearly, his temperament contributed most to his demise. The simmering ill-will boiled over in 1923 and 1924 when he began to stay at Langley for weeks at a time to run the Variable Density Tunnel. During those years, first the lab's chief physicist, Frederick Norton, and then his successor, David Baco, resigned in protest over Munk's high-handed conduct.

The finale to this affair occurred in 1926 when George Lewis made one of the worst mistakes of his career: transferring Munk to Langley and naming him the chief of the Aerodynamics Division. Munk now ran the wind tunnels, flight research, and analytical studies, and assumed a rank second only to Henry Reid in the lab hierarchy. Within a year, the Langley staff mutinied against Munk. All four of his section chiefs tendered their resignations. Faced with rebellion, Reid had to act, even though he had only been on the job for a year. He relieved Munk of his duties and made him his adviser. Although Lewis offered to take him back in Washington, Munk resigned, his pride wounded.

This event ended a period in NACA history that was divisive yet at the same time productive, all the more because it occurred almost at the start of the organization's existence. Max Munk never again achieved the same eminence he enjoyed at the NACA, sustaining himself during the rest of his long life by part-time employment, such as teaching mechanical engineering at Catholic University in Washington, D.C. But the NACA lost just as much: a researcher capable of seminal discoveries, a man of international reputation, and a scholar whose work burnished the young institution that employed him.

Tragic as the Max Munk affair had been, Lewis and Reid hired many other talented people at Langley. Once the lab's star began to rise with the Variable Density Tunnel, other researchers found their way to the NACA, and

Max M. Munk

The son of German Jewish parents from Hamburg who intended him for rabbinical study, Max Michael Munk (born in 1890) instead entered the Hanover Polytechnical School. After earning an engineering degree, he enrolled at Göttingen University and studied with Professor Ludwig Prandtl, the world's leading authority on fluid dynamics. Munk took two doctorates in 1917, in engineering and in physics. One of Prandtl's finest students, Munk decided to emigrate to the United States. Prandtl broached the subject with an American acquaintance, Dr. Jerome Hunsaker, a former professor at Massachusetts Institute of Technology (MIT) and subsequently chief of the Aircraft Division of the Navy's Bureau of Construction and Repair. He had visited Göttingen on a tour of European aeronautical facilities and appreciated the German approach to research. Hunsaker, in turn, contacted Joseph Ames, who secured Munk's employment with the NACA. During his first six years, Max Munk acted, in effect, as an in-house theorist, serving as George Lewis's technical assistant in Washington, D.C. He arrived in Washington in 1920 at the age of thirty, already a man of distinction in his field. His pioneering research culminated in the publication of more than forty papers by the NACA, including such diverse subjects as airflow around airships, centers of aerodynamic pressure on a number of aerodynamic shapes, and airfoil theory. A difficult temperament, Munk resigned from the NACA after only seven years and died after a long life of comparative obscurity in 1986.

←···· German émigré Dr. Max Michael Munk gave the early NACA its first major technical successes—especially the concept of the Variable Density Wind Tunnel—but also its worst bureaucratic crisis. Embroiled in a mutiny by his subordinates at Langley, Munk resigned at the age of thirty-six. Tragically, although he lived a very long life, in his remaining years he achieved almost nothing professionally.

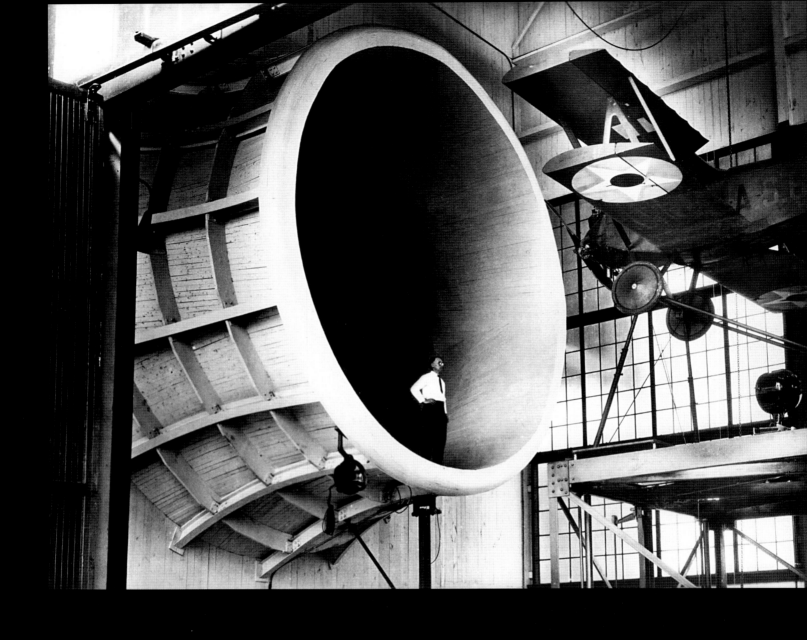

[Opposite, top and bottom far left] *If the Variable Density Tunnel represented the "state of the art" in tunnel design during the early 1920s, Langley scored a second triumph later in the decade with the huge Propeller Research Tunnel. Big enough to contain full-sized portions of aircraft, it occasioned seminal cowling research under section chief Fred E. Weick. In the top picture, Max Munk's successor Elton Miller stands in the exhaust cone of the Propeller Research Tunnel in a 1927 photograph. At the far lower left, technicians make adjustments.*

[Opposite, bottom right] *A Langley technician installs an airfoil for testing in the Variable Density Tunnel in early 1925. Engineers found that the data acquired from this apparatus corresponded closely to that derived from actual flights in the atmosphere. Such accuracy was unusual for ground testing during the 1920s, and was an important reason why the NACA won growing international recognition for its aeronautical research.*

[Above and right] *A mock fuselage mounted in the Propeller Research Tunnel shows the scale of the huge machine. Experiments in the tunnel resulted in a set of cowlings that covered radial engines in order to reduce aerodynamic drag, yet at the same time enabled enough air to reach the powerplants for cooling. This chart, produced for the 1930 NACA manufacturers' conference, illustrates the advantages of various cowlings to improve lift and diminish drag.*

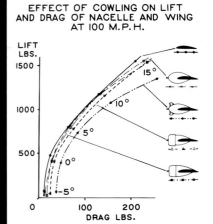

EFFECT OF COWLING ON LIFT AND DRAG OF NACELLE AND WING AT 100 M.P.H.

4430

subsequently discovered ways to raise their own stature,
as well as that of the laboratory. Fred E. Weick
represented one of Langley's most talented recruits.

Weick's youth and inexperience did not prevent George
Lewis and Henry Reid from bestowing heavy
responsibilities on him. Lewis and the NACA Executive
Committee undertook the design and construction of the
immense Propeller Research Tunnel because of repeated
discrepancies between full-scale propeller data and that
derived from models. This machine, with a capacious 20
foot (6m) diameter throat, enabled researchers to test
full-sized propellers, a far more efficient and speedy
method than flights in the atmosphere. It also permitted
experiments on other full-scale aircraft parts, such as
fuselages, tail sections, landing gear, and large airfoil
models. But, however gargantuan in size by contemporary
standards, the Propeller Research Tunnel dwarfed the
other Langley machines even more in terms of the power
required to animate it. It required 1000 horsepower to
operate at the required 100 miles (161 km) per hour, ten

DURING 1929, EVERY AIRCRAFT THAT
BROKE A SPEED RECORD IN THE U.S.
FLEW WITH A RADIAL ENGINE, COVERED
BY AN NACA COWLING.

times the energy needed by the original 5 Foot Atmospheric Wind Tunnel and more electricity than the cities of Hampton or Newport News had available. To compensate, the NACA took the by now familiar step of borrowing from its military counterparts. Usually these loans involved aircraft, but in this case the U.S. Navy provided two 1000 horsepower diesel submarine engines. Thus, until 1933 when the electrical grid around Langley could support it, the big tunnel operated with Navy hand-me-down power generating equipment.

During those early years, the Propeller Tunnel, even if powered in a makeshift fashion, enabled the Langley engineering staff to make some pivotal aeronautical discoveries. One such project probably originated with aircraft designers at the Navy's Bureau of Aeronautics (where, coincidentally, Fred Weick had worked when Lewis met him). Due to the need for carrier-based landings, Navy officials decided after World War I to rely on the lightweight, air cooled radial engine, rather than the more traditional water cooled power plants favored in Army

aviation circles. The choice made sense; jarring touchdowns and inadequate space for spare parts argued persuasively for the lighter, simpler, and more durable radials. But these advantages came at a price. Since air flow moderated the engine's temperature, it needed to be uncovered, resulting in a loss of thrust due to heavy aerodynamic drag. However, some in the Bureau of Aeronautics wondered whether it might be possible to cover radial engines in flight, yet not inhibit cooling. George Lewis, a propulsion man himself, decided to give his "boys" at Langley the task of finding out.

Weick, and his small group of Propeller Tunnel technicians and engineers, initiated the project in May 1926. The concept of shielding radial engines with custom built cowlings had also been suggested that same month by aircraft designers during the first annual NACA manufacturers' conference in Hampton. These meetings— part technical expo and part public relations show— included a tour by Lewis and Reid of the laboratory and informative lectures about ongoing Langley work for

↙ ↓ *Metal fabricators in the Langley shops painstakingly fashioned the cowlings according to the specifications of Weick and his team. They were tested in wind tunnels, but the cowlings also underwent extensive flight in the atmosphere to determine their effectiveness. A Curtiss AT-5A on loan from the armed forces provided the flying testbed. Shown here without and with the famed Number 10 cowling, the aircraft proved to the NACA pilots that a big increase in airspeed occurred with the application of Weick's new device.*

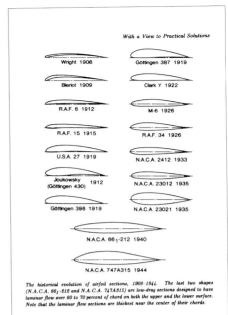

The historical evolution of airfoil sections, 1908–1944. The last two shapes (N.A.C.A. 66₁-212 and N.A.C.A. 747A315) are low-drag sections designed to have laminar flow over 60 to 70 percent of chord on both the upper and the lower surface. Note that the laminar flow sections are thickest near the center of their chords.

↑ The efficiency to be gained by the NACA cowlings soon became known throughout the world, both in commercial and in military aeronautics circles. Its discovery represented yet another achievement that won distinction for the Langley Aeronautical Laboratory, not yet ten years old when President Herbert Hoover (at left of table) presented Joseph Ames (right of table) with the famed Collier Trophy on the lawn of the White House in 1929.

↗ A chart illustrating the development of airfoil sections up to 1944. The NACA played a pivotal role in their evolution, beginning in 1923 in Langley's Variable Density Wind Tunnel. Starting with an intuition by Max Munk—that airfoil research should proceed by "fitting" the mathematics to existing wing sections rather than waiting for mathematicians to discover improved models—NACA scientists and engineers produced a series of efficient airfoils for a variety of purposes.

the visiting industry leaders, politicians, officials representing the Washington, D.C. bureaucracy, and university researchers.

The high point of the 1926 conference occurred when technicians animated the still unfinished Propeller Tunnel, an impressive and noisy demonstration that apparently prompted some taking the tour to suggest it as a research platform for the cowlings. The NACA Executive Committee acted quickly on this suggestion, initiating Research Authorization (RA) 172. (Research Authorizations gave specific approval and guidance to Langley engineers and scientists to launch their investigations. In reality, however, Henry Reid and George Lewis gave their staff wide latitude in satisfying the spirit of the RAs.) This particular Research Authorization empowered both tunnel tests and actual flight in aircraft. Weick took command of the project.

Fred Weick's first insight in the cowling research certainly proved its value. He decided to approach the problem of aerodynamic drag and engine cooling using a venerable technique that had only recently been recognized by the American engineering profession. Called experimental parameter variation, it enabled researchers to avoid the morass of causation and instead pursue a series of experiments that systematically yielded data based on the broad parameters of the cowling's operational range. Eventually, the empirical data compiled by these repetitive tests would reveal a pattern that illuminated the underlying reasons why some designs worked and others failed. Thus, Weick began with an engine nacelle (a housing for the engine and propeller)

that approximated a familiar shape: the nose end of an airship, into which air flowed through the center of the nose. He then fashioned ten different cowlings to cover the nacelle, gradually running the spectrum between one that substantially enclosed it and one that left it mainly bare. At the extreme end of the parameter he added an uncowled radial engine, which (intuitively) offered greatest cooling but greatest drag; and at the other extreme added a cowling that totally covered the engine, presumably causing maximum aerodynamic efficiency but minimum cooling.

Weick then mounted sequentially each of the twelve variations in the Propeller Tunnel (which became operational in 1927) with the goal of finding a cowling that provided the maximum engine enclosure, yet enabled as much cooling as one with no cowling at all. This optimistic outcome seemed attainable until the fully covered cowling (known as Number 10) went onto the nacelle and caused unacceptably high temperatures in some of the engine's cylinders. But, after considerable experimentation, the engineers discovered that carefully modifying the cowling's air inlet and exit paths and adding internal guide vanes, resulted in the cowling Weick wanted—one that cooled reliably and adequately yet also reduced drag.

One important surprise came to light. The tunnel research suggested not just an increase in aerodynamic efficiency with Cowling Number 10, but a vast improvement. According to the findings, the cowling reduced drag by a factor approaching three. Rather than await confirmation by flight research, the NACA leaders decided to make this pivotal news available to the aircraft industry at once. In November 1928, Fred Weick published NACA Technical Note 301 announcing the results. But one more bombshell remained. After the release of the Technical Note, Langley researchers outfitted a borrowed Army Air Service Curtiss AT-5A aircraft equipped with the same Wright Whirlwind J-5 engine used in the Propeller Tunnel. Unaided, the AT-5A flew at 118 miles (190 km) per hour. Technicians then installed Cowling Number 10. With it, the plane achieved a speed of 137, and at times 138, miles per hour (220.5–222 km per hour).

This achievement placed Langley and the NACA in the forefront of worldwide aeronautical research. Far more even than the esteem accorded the NACA by the path-breaking Variable Density Wind Tunnel, the cowling discovery—demonstrated in the spanking new Propeller Research Tunnel—made the NACA a name to be reckoned with in international aeronautics. The agency announced to industry that for the cost of about $25 per cowling, it stood to conserve about $5,000,000 in fuel annually, more than the total budget of the NACA since its founding, and a massive saving when compared to the Propeller Tunnel's initial cost of $291,000. The practical value of the discovery materialized almost immediately. During 1929, every aircraft that broke a speed record in the U.S. flew with a radial engine, covered by an NACA cowling. That same year, in front of the White House, President Herbert Hoover presented NACA chairman Joseph Ames with the Collier Trophy, awarded annually by the National Aeronautic Association for the year's outstanding aeronautical achievement.

During its early period, the NACA gained fame from its wind tunnel research, a tradition begun by Max Munk, continued at Langley in groundbreaking airfoil research by men such as Eastman Jacobs and the Norwegian Theodore

Theodorsen, and pursued by Fred Weick and his team with cowlings. But the other half of Langley research made equally important contributions to aeronautical knowledge. On the Langley flight line and in the hangars, engineers, pilots, mechanics, and machinists were engaged in seminal flight research work. Experiments aboard full-sized aircraft had in fact occurred before the first wind tunnel went into service, and indeed constituted Langley's very first cluster of experiments. Even as work gangs drained the marches and ripped tree stumps from the ground, pilots flew overhead and gathered data in two Curtiss JN4H Jennies, borrowed from the Army Air Service. Ultimately, Langley's chief physicist Edward P. Norton—a young MIT aeronautics professor of extraordinary talents—initiated these investigations in order to gauge the accuracy of the 5 foot (1.5m) wind tunnel still under construction. If the data gathered from the Jennies in flight matched the data generated later by a model in the tunnel, the tunnel's accuracy could be verified. As it turned out, the new tunnel proved to be only partially reliable.

In future experiments, the NACA's engineers would fly full-scale aircraft routinely to test the accuracy of the many tunnels that eventually rose on the Langley "campus" (as it came to be called). But flight research also played an important independent role in the course of aeronautical history, and the NACA used it repeatedly for this purpose. Perhaps the most influential example of such a project occurred during the mid- to late 1920s, a time when increasingly powerful military aircraft were developing the disconcerting habit of disintegrating during combat maneuvers. Army Air Service Captain James H. Doolittle, looking for a thesis subject for a master's degree, received an assignment from the

⬋ *By the late 1920s, the NACA's reputation for research attracted famous visitors. Legendary aviatrix Amelia Earhart is pictured here with Engineer-in-Charge Henry Reid (second from right, wearing glasses), Langley pilot Thomas Carroll (fifth from left, with moustache), and Propeller Research Tunnel chief Fred Weick (third from right). Earhart toured the laboratory in November 1928, fresh from being the first woman to cross the Atlantic Ocean in an aircraft (as a passenger, not a pilot). On this visit to Hampton, part of her fur coat became ingested in the 11 Inch High-Speed Wind Tunnel.*

⬇ *The year before Amelia Earhart met with Henry Reid and his staff, Charles Lindbergh—less than two weeks after his triumphant solo voyage over the Atlantic—came to Hampton, in June 1927. The photographer captured him in the front cockpit, Fred Weick in the rear. In later years, Lindbergh sat on two NACA panels, the prestigious Main Committee from 1931 to 1939, and (as chair) the Special Committee on Aeronautical Research Facilities (in 1939).*

generals at Wright Field's Engineering Division in Dayton, Ohio, to study the problem of pressure distribution on aircraft flying extreme maneuvers.

In 1924, acting both as pilot and chief engineer, Doolittle piloted a Boeing PW-9 (instrumented with an NACA-type accelerometer) through a series of hair-raising dives, pull-ups, rolls, and other hazardous maneuvers in order to determine the effects on the aircraft. He flew just to the point of failure and undoubtedly risked his own life frequently, often experiencing force six or seven times that of gravity. Inspections of the PW-9 after his flights confirmed the anecdotal evidence associated with the recent crashes: structural members had been severely affected, and the data from the accelerometer proved the stresses conclusively. But the Air Service needed to know more: Exactly what caused the crashes? Could aircraft be designed to withstand the forces that caused these catastrophes? Until the officers at Wright Field had answers, no combat aircraft could be considered safe.

They therefore called upon George Lewis and the NACA

for assistance. The project began in 1926, just as Henry Reid became Langley's Engineer-in-Charge. By sheer coincidence of timing, the inexperienced Reid found himself not only in charge of the decisive cowling tests but also responsible for the NACA's first major flight research project. To undertake the pressure distribution tests a new airplane needed to be purchased, something the NACA had not yet done in its ten year history. Until now, it had conducted research by borrowing Air Service and Navy aircraft, or by taking possession of military cast-offs (a practice that continues to the present day). But in this instance, Lewis and Reid decided that repeating the highly risky flights of Doolittle demanded a vehicle specially designed to endure the repeated stresses imposed on it. They ordered a Boeing PW-9—the same type as that chosen by Doolittle, a sturdy, no-nonsense plane with big wheels and plenty of wing bracing—and asked the manufacturer to strengthen the tail and fuselage according to NACA specifications. Its delivery launched NACA Research Authorization 138, "Investigation of Pressure Distribution and Accelerations

↖ ↑ *Two Langley mechanics measure the wing ordinates of a Curtiss JN4H Jenny in 1921. Early flight research aboard hand-me-down Army Air Corps vehicles, such as this one, not only tested and calibrated the accuracy of Langley's growing family of wind tunnels but also collected flight data unavailable by any other means. The Langley hangar, seen here around 1933, sheltered these and future research aircraft (the vast majority of which arrived at Langley by loan agreements from the military services, not by purchase).*

on Pursuit-Type Airplanes," authorized by Joseph Ames two years earlier.

During their first decade of existence, the Langley flight research group evolved a simple but fundamental approach to flight research: instrument the experimental aircraft thoroughly; collect the required data in precise, predictable increments; and never jeopardize the lives of the pilots. Accordingly, the researchers decided to conduct some of the most dangerous parts of the project in the Langley tunnels while they awaited the arrival of the special PW-9 from Boeing. To facilitate the tests, a young Langley engineer in charge of the Atmospheric Wind Tunnel named Elliott Reid (unrelated to Henry) devised an apparatus by which model airfoils moved freely— pitching up or down—in a stream of air, enabling his staff to begin to understand how wings behaved as they responded to the pressures experienced in flight. Gradually, the team discovered the point at which the pitching motion of the wings failed to provide lift, knowledge that could then be correlated with data from actual flight in the atmosphere and subsequently

become part of the guidelines for safe flight under high air pressure.

Before that realization, enormous amounts of work needed to be done by the Langley test pilots. To guide the flight research on pressure distribution, Henry Reid selected yet another young engineer. Just twenty-eight years old at the start of the pressure distribution research, Richard V. Rhode proved to be as effective as Fred Weick had been with the cowlings. But Rhode's assignment entailed worries that Weick's did not. Day after day, the Langley pilots put the PW-9 through tortuous maneuvers in order to retrieve worthwhile data, a chancy proposition despite the NACA desire for safety. Over time, Rhode's group compared the existing aircraft design criteria with levels of safety dictated by the new research. On occasion, the discrepancy could be shocking due to the onrush of technology and the rapid rise in the capacities of military aircraft. For instance, the loading on the leading edges of the wings required an entirely new standard. As recently as the early 1920s, the NACA had flown a Thomas Morse aircraft which pulled out of dives

James H. Doolittle

James H. Doolittle possessed unique personal and professional qualifications for the daring pressure distribution tests. Just twenty-eight years old in 1924, he had already experienced much of life. Doolittle grew up in Alameda, California, lived in Alaska, flew as a cadet in World War I, won a commission in 1918, graduated from the University of California, Berkeley, and in 1922 made the first cross-country flight using primitive instrumentation. He won the Schneider, Bendix, and Thompson air races. In 1925 he earned a Doctorate of Science from MIT. Doolittle then continued a life of notable achievement, becoming a corporate executive with Shell Oil after leaving Army Air Corps active duty in 1930. Re-joining the service at the start of World War II, Doolittle is perhaps best remembered as the architect and leader of the successful bombing raid on the Japanese mainland in April 1942 subsequent to the attack on Pearl Harbor. Doolittle retired from the Army Air Forces after World War II with the rank of lieutenant general. During the mid-1950s he served as the last chairman of the National Advisory Committee for Aeronautics. Doolittle won elevation to four-star rank from President Reagan in 1985 and died in 1993 at ninety-seven years of age.

at a maximum speed of 150 miles (241 km) per hour, yielding pressures of 200 pounds per square foot (970 kg per square meter) on the leading edges of its wings. Now, the PW-9 achieved 186 miles (299 km) per hour in dives, but yielded an alarming 400 pounds per square foot of pressure at its leading edges as it pulled up. Clearly, to avoid catastrophe, the aircraft rolling off the assembly lines at the end of the 1920s required strengthened leading edges. Far more disturbing for aircraft designers, however, was the discovery by Langley researchers that in the 1.5 seconds it took a pilot to pull up from a dive at 186 miles per hour, pressure on the aircraft rose from 1 g to 9 gs *in half a second*: an almost instantaneous rise in pressure.

The publication of the results of the pressure distribution program won the NACA laurels equal to those garnered by the cowling research. For the first time, aircraft designers could build their vehicles with a sound, verifiable data base, augmented by detailed pilot impressions that showed the limits of safe design. It also won acclaim for Richard Rhode. Since its inception, the NACA had propagated its research findings by technical reports, memoranda, and notes, published only after stringent internal review and designed not just for the American audience but for worldwide consumption. As the principal author, Rhode became well known for his technical achievement, as well as for his customary confidence.

---→ ---→ ↗ *Speed pilot, military aviator, aeronautical researcher, corporate executive, leader of the World War II Doolittle Raiders, and the last chairman of the NACA, James H. Doolittle appeared at Langley in 1928 with his Curtiss Racer, the plane that had flown him to victory in the 1925 Schneider Cup race. Earlier in the decade Doolittle had undertaken a series of hair-raising flights aboard a Boeing PW-9 military pursuit aircraft, which in turn launched the NACA on a highly productive research project that uncovered some of the mysteries of pressure distribution on aircraft structures. It underwent tests both in the Langley Full-Scale Tunnel and in the skies as Langley pilots conducted a long series of flights in the PW-9 (shown here). Unlike Doolittle's plane, this one had*

been built to order for the NACA by the Boeing Company, which had strengthened its wings, tail surfaces, and fuselage in order to withstand the punishing pressure distribution tests. It represented a rarity: an aircraft purchased, not borrowed, by the NACA.

These photographs illustrate some of the varied workforce at Langley roughly at the time of the pressure distribution tests (late 1920s to early 1930s). At the top of the hierarchy, Henry Reid (top picture, third from left) greets two guests to the laboratory, accompanied by test pilot Thomas Carroll (far left) in 1928. Despite the demands of a more formal age, Langley employees dressed informally during the hot, oppressive Hampton summers, as the picture of engineers in shorts demonstrates. Pictured eighth and tenth from left are aerodynamicists Eastman Jacobs and John Stack, respectively; fourteenth is test pilot Melvin Gough. The image of the administrative offices (bottom left) show the surprisingly spacious and orderly working conditions of the non-technical staff. Finally, at the end of 1928, all employees at the laboratory celebrated Christmas with the annual party at the Langley boat house (bottom right).

The first public glimpse of the pressure distribution research occurred in January 1929 in a preliminary paper for the Wright Field Experimental Engineering Section entitled "A Danger in Maneuvering Airplanes of Similar Type." Much to the anger and disappointment of Rhode and his Langley cohorts, the Air Corps—the patron of this project—actually rebuked the NACA's report for tailoring it to pilots, rather than to the nation's structural engineers, and for exaggerating the inadequacy of existing design standards. But Rhode—who was not yet thirty—held his ground, both for himself and for his institution. He argued that pilots should be the first to be forewarned, in order to safeguard their lives, and that since the present generation of combat aircraft produced from 7 to 9 gs in ground attack, the dangers could not be denied. Rarely did the Air Service or its successors take this approach to NACA research again. Indeed, requests to Langley mounted from the military services, including the Navy and Marine Corps. But more than that, American industries and academics, as well as foreign governments and manufacturers, beat a path to the NACA's door. The research even attracted the attention of journalists, and in 1931 the magazine *Aviation* featured an article by Rhode explaining his conclusions to a lay audience.

Fresh from the triumphs of cowling and pressure distribution research, the Langley engineers looked for new projects for themselves and for the NACA. By the mid-1930s, the pursuit of speed began to entice many. Two of Langley's most able minds led the chase. Bold and imaginative, Eastman Jacobs, a graduate of the University of California who joined the NACA in 1925, devoted much of his career at Langley to developing a family of NACA airfoils suited to specific aeronautical purposes, a list that eventually found almost universal use around the world. John Stack, a gritty, straight-talking outdoorsman, came to Langley fresh out of MIT in 1928 and in a short time became one of the laboratory's most respected and open-minded researchers. During the 1930s, Stack worked for Jacobs, who headed the Variable Density Tunnel group at Langley. But in 1941, Stack became chief of the newly formed Compressibility Research Division.

Compressibility—the turbulent aerodynamic effects encountered by aircraft approaching the speed of sound—had first been investigated during the early 1920s by two physicists at the National Bureau of Standards, Drs. Hugh L. Dryden (the future successor of George W. Lewis) and Lyman J. Briggs (later the director of the Bureau of Standards). Their preliminary research illuminated the basic character of transonic flight (that is, flight in an aerodynamically unstable region just below and above the speed of sound). Jacobs and Stack collaborated on the problem experimentally during the

←···· *Newly installed director of the NACA, Hugh L. Dryden, far left, visits the Langley laboratory accompanied by John Victory (center) in September 1947. Greeting them is Henry Reid. Although employed by the National Bureau of Standards from 1918 to 1947, Dryden served on NACA committees from the 1920s onwards, and his appointment as George W. Lewis's successor surprised few. Dryden's early, pioneering publications on high-speed flight proved to be the keynote (but certainly not the sole research interest) of his career, informing the agenda of the NACA when he became its leader and preparing him well for his final role as the first Deputy Administrator of NASA.*

↑ ⟶ By the 1930s and 1940s, life at Langley had assumed the feel and look of a college campus. Among the leading men on this campus, John Stack (right), the lab's Chief of Compressibility, not only became an early adherent of supersonic flight, but probably did more than any other individual to hasten its development by proposing research aircraft to demonstrate flight up to and over the so-called sonic barrier. The atmosphere and infrastructure at Langley lent itself to technological leaps such as these. In the model shop, for instance, a cadre of fine young technicians produced handmade, highly accurate aircraft miniatures for spin research.

↗ The annual Engineering Conferences, begun in 1926, became defining events in the history of the laboratory, eventually attended by hundreds of industrialists, politicians, bureaucrats, academics, and military officials eager to see the latest research at the nation's leading aeronautical research center. Here, the attendees of the 1934 conference pack under the massive roof of the full-scale wind tunnel, the only place on campus big enough to accommodate them all.

late 1920s and early 1930s, building an 11 inch (29 cm) high-speed wind tunnel (its airflow derived from the Variable Density Tunnel), and actually observed compressibility phenomena using schlieren photography (which enabled pictures to be taken of supersonic flow). Indeed, as early as 1934, in an article published by the *Journal of the Aeronautical Sciences,* Stack actually conceived of a transonic airplane capable of speeds of 566 miles (910 km) per hour. But he also realized that the transonic region needed to be understood more fully before aircraft designers could fabricate machines capable of penetrating it safely.

Stack and Jacobs consulted the Langley wind tunnels for the desired data. But to their puzzlement and frustration, they encountered a strange "choking" or turbulent effect above Mach 0.7 that continued just past supersonic speeds and prevented worthwhile data from being collected. By 1941, Stack had concluded that for the time being at least, wind tunnels could not unmask the secrets of the transonic region. He therefore decided to

press the case with George Lewis for fabricating a full-scale aircraft to demonstrate flight through the transonic zone. Ever cautious, Lewis resorted to an administrative technique he often adopted. Instead of giving Stack permission to pursue the project openly under a Research Authorization, Lewis told him to make discreet inquiries and work up some "back of envelope" estimates for a transonic airplane. Stack asked several of his lieutenants to join the project and in summer 1943 they unveiled their design: a turbojet aircraft capable of Mach 0.8 to Mach 1.0. Stack then organized a meeting at Langley in the spring of 1944 between NACA engineers and representatives of the Army and Navy, but the two services bickered over design details.

Meantime, just as stalemate gripped both high speed wind tunnel research and Stack's proposal for a transonic aircraft, World War II combat aircraft reached a dangerous performance wall of their own. Although the problem of wind tunnel choking eventually fell to researchers at Langley who solved the problem by placing slots in the

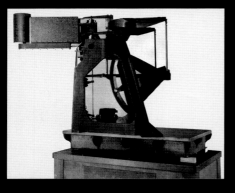

Beginning in the late 1920s, the pride of Langley became its unmatched family of wind tunnels. Some of the many that rose on the Hampton campus included the 11 Inch High-Speed Tunnel, filled with compressed air bled from Munk's Variable Density Tunnel. Shown here (top left) is its automatic recording balance which measured the lift, drag, and pitching of models at high speeds. A year after the 11 Inch opened, in 1929 the 5 Foot Vertical Tunnel enabled researchers to test spin recovery properties of aircraft without risking crews or airplanes (top centre). The following year, the 7 by 10 foot (2 by 3 m) tunnel started operations, designed for general purposes as a replacement for the original Langley 5 Foot (top right). Then, in May 1931, the behemoth 30 by 60 foot (9 by 18 m) open-

throat Full-Scale Tunnel became available to researchers, capable of a maximum airspeed of 118 miles (190 km) per hour, and costing $900,000. It is shown here testing two full-sized propellers, accompanied by a man holding up a piece of the scale model tunnel that preceded construction of the real one (opposite, top and left). Just six years after the 11 Inch became operational, in 1934 John Stack, Eastman Jacobs, and other Langley engineers devoted to transonic flight began to use the 24 Inch High-Speed Tunnel, equipped with schlieren photography that illustrated the motion of shock waves (above). In June 1939, the 19 Foot Pressure Tunnel opened, one of the first to combine high pressure and large size in order to investigate the persistent differences between scale model findings and those of

full-sized aircraft (above left). Early in World War II, NACA engineers curious about the maneuvering performance of combat aircraft had at their disposal the Stability Tunnel, the control room of which is shown here (left). Finally, the 16 Foot High-Speed Tunnel (opposite, right), completed just as the U.S. entered World War II, caused intense frustration among Stack and his staff. Between 1941 and 1943 they discovered severe "choking," or distortion in the new machine above a speed of Mach 0.7, resulting in a termination of the tests and a realization by Stack that at the present state of the art, he needed to find other means of crossing the transonic frontier.

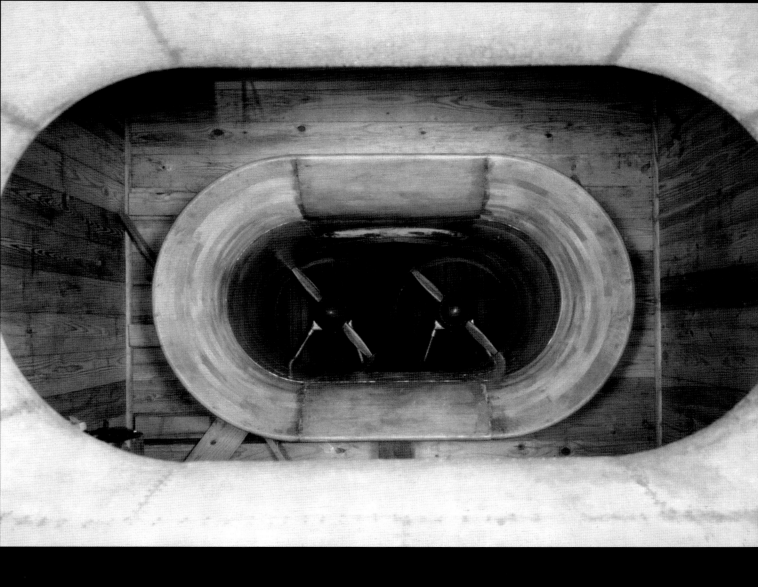

↑ ↗ *Frustrated by the inconclusive data produced at transonic speeds in the 16 Foot Wind Tunnel, Stack and others at Langley turned to actual flight vehicles. They decided to ask the Langley pilots to conduct a risky series of flights. Using the fast and powerful front-line American fighter aircraft of the day—the P-51 Mustang, shown here in a partial front view, a side view (the D-model), and mounted in the Full Scale Tunnel—the flyers risked their lives as they sent the aircraft*

into steep dives again and again in order to attain the maximum airspeed. The resulting data convinced Stack of the need for custom made research aircraft capable of penetrating the so-called sound barrier.

throats of tunnel test sections, the impatient and hard-driving Stack wanted answers to the transonic dilemma now, not later. He received help from another youthful Langley researcher named Robert Gilruth, the chief of the lab's flight research section. Gilruth conceived the idea of mounting a model airfoil vertically on a P-51D Mustang's wing, instrumenting it thoroughly, and then instructing the Langley pilots to fly it in sharp dives. As the planes plunged, they flew at a speed of about Mach 0.81; but the airflow around the model actually jumped as high as Mach 1.4, enabling measurements to be taken of the full range of transonic behavior. As data from these perilous

flights materialized, it offered some clues about what might lie ahead for a true transonic vehicle. Consequently, Stack pressed forward with his initiative for a full-scale high-speed aircraft. But meetings with the Army Air Forces in May and December 1944 showed that the service had its own transonic objectives, not necessarily compatible with those of the NACA. The generals wanted a supersonic aircraft capable of Mach 1.2 flight, powered not by a turbojet but by a rocket. Ultimately, the Air Forces picked Bell Aircraft of New York—manufacturer of the first American jet (the XP-59 Airacomet)—to build their supersonic plane.

Stack, meanwhile, called on friends in the Navy Bureau of Aeronautics to act as a counterweight. Always keen to differentiate themselves from the Army, the Navy Bureau of Aeronautics accepted Stack's concept of transonic research and provided the funds for the project. In contrast to the Army Air Forces, which had contracted with Bell in the past, the Navy and the NACA had a greater familiarity with Douglas Aircraft because of the California firm's past work on naval aircraft. Thus, the nation found itself with not one but two high-speed aircraft programs, the Bell Experimental Sonic (XS)-1 —more often known as the X-1—and the Douglas D-558.

 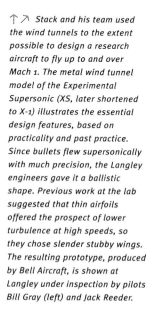

Stack and his team used the wind tunnels to the extent possible to design a research aircraft to fly up to and over Mach 1. The metal wind tunnel model of the Experimental Supersonic (XS, later shortened to X-1) illustrates the essential design features, based on practicality and past practice. Since bullets flew supersonically with much precision, the Langley engineers gave it a ballistic shape. Previous work at the lab suggested that thin airfoils offered the prospect of lower turbulence at high speeds, so they chose slender stubby wings. The resulting prototype, produced by Bell Aircraft, is shown at Langley under inspection by pilots Bill Gray (left) and Jack Reeder.

But in reality, the two projects evolved in close cooperation, mainly because of intense NACA participation in both. Stack and his associates played an intimate role in the design of the X-1. Indeed, the Army contracted with Bell to make two X-1s: one for the Army's purposes, the other to be transferred to the NACA for research. Not only did the instrumentation package consist almost exclusively of NACA equipment and design, but Bell also accepted the NACA's recommendations to make the wings thin for improved transonic flight, and straight, rather than swept back. (The Army's X-1-1 would have the thinner wings; the NACA's X-1-2 would be a little thicker, in order to make comparisons of their flight characteristics.) Still, even though NACA staff found much about the X-1 to dislike—for instance, rocket propulsion offered a potential for interference with the readings of the instrumentation that turbojets did not—Stack and his lieutenants gave their full cooperation to the Army and Bell, even conducting subsonic wind tunnel tests upon

request. But Stack never deviated from one point: both the Bell X-1 and the Douglas D-558 existed solely to gather data about compressibility that was not available by any other means.

Because the Army and Bell decided to launch the X-1s from a bomber aircraft rather than the ground, a larger air strip than Langley's needed to be found. They chose Pinecastle Army Air Field near Orlando, Florida, because of its immense (10,000 foot/3048 m) runway. The NACA sent a small contingent to Orlando to instrument the X-1 prior to its first flights, scheduled for January 1946.

Walter C. Williams headed the NACA delegation. Just twenty-seven years old, this forceful and tough-minded stability and control engineer had worked closely with John Stack and looked forward to the challenge. The Pinecastle tests occurred during January, February, and March and, according to Bell's test pilot Jack Woolams, the bullet-shaped aircraft demonstrated good handling qualities as it glided from the B-29 mothership to the airstrip below. But the dense forestation around Pinecastle proved problematical during landings, and for that reason Bell and Army officials decided to find a different place to run the later rocket-powered flights. They sent Woolams on a cross-country search for the best facility. He had already seen one spot that looked promising when he had been involved in Bell's X-59 project. About 100 miles (160 km) northeast of Los Angeles, in the desolate eastern Mojave Desert, lay the immense but sparsely populated Muroc Army Air Field, situated at the edge of Rogers Dry Lake, a hard, white, flat plateau 12.5 miles (20 km) long and 5 miles (8 km) wide at the maximum extent. Woolams liked it from the start, but ruled it out for the glide tests because rainwater had accumulated on the lakebed during the winter of 1945. Now, the Bell pilot recommended Muroc. It had many advantages: open skies, fine weather, privacy due to isolation, and the vastness of Rogers Dry Lake for emergency landings. The Army brass at Wright Field

The first X-1 flight tests—unpowered glides from the belly of a bomber—occurred at Pinecastle Army Air Field in Orlando, Florida. Looking for more open terrain for the powered flights, the Army Air Forces selected Muroc Army Air Field in the desolate western Mojave Desert of Southern California. It was here that the first American jet, the Bell X-59A Airacomet, had undergone flight tests during World War II. In September 1946, a contingent from the Langley Laboratory arrived at Muroc to conduct the X-1 supersonic experiments. Called the Muroc Flight Test Unit, some of its staff (top left) confer about the next X-1 flight. The NACA's leader at Muroc, Walt Williams (fourth from left, holding the picture) is flanked on his right by Captain Charles "Chuck" Yeager. To fly the missions, the X-1 technicians needed to mount the research airplane onto the enormous B-29 mothership, a complicated and time consuming process in which they lowered the X-1 into a loading pit, positioned the B-29 over it, and hung the little vehicle by hooks and straps.

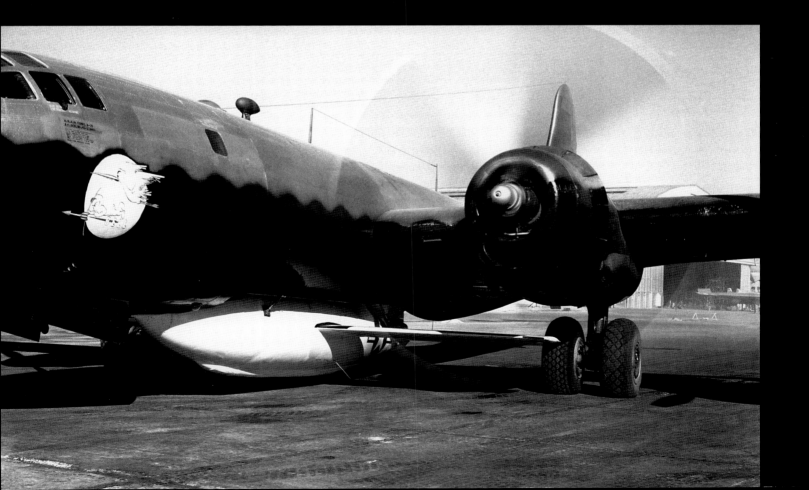

↑ *Even while the flight research on the X-1 progressed, Langley researchers continued to subject models to wind tunnel scrutiny. The 16 Foot High-Speed Tunnel effectively recorded transonic data once it had been modified with a slotted-throat test section in 1950.*

↗ ---→ *The real test of the X-1 occurred in the skies over Muroc, but much of the flight research also happened on the ground. Here, instrumentation experts lay out their equipment on a cart prior to packing it into the waiting X-1. These devices proved indispensable in recording the aircraft's behavior in flight. The stuffed X-1 cockpit reflected the pilot's main challenge once airborne: to control the airplane while never taking his eyes off of the gauges.*

agreed; Muroc became the home of the X-1s.

The flights of the Bell X-1s at Muroc brought together an unlikely set of characters. But among the organizations that contributed during 1946 and 1947 to the world's first supersonic flight, the NACA faced the biggest hurdles. The Army Air Forces, the Navy, even the Marines had been in the area for some time, as had Bell Aircraft for the XP-59 Airacomet flights. In contrast, the NACA had never operated on Muroc and needed not only to set up an infrastructure—hangar space, offices, and housing—but at the same time prepare for perhaps the most important flights since those of the Wright brothers. But the Langley team handled the problem gracefully. They put the task in the hands of Melvin Gough, the laboratory's chief of the Flight Research Division and a pilot of twenty years' standing. It fell to Gough and John Stack to choose the NACA's Muroc leader, and both agreed that the formidable Walt Williams had the toughness and the will to deflect challenges that might arise from Muroc base officials, Army Air Forces technical officials, or representatives of Bell. Williams arrived on September 30, 1946, with two other engineers, and together with two earlier arrivals they formed the NACA Muroc Flight Test Unit.

Williams struggled to accommodate the needs of his fledgling staff, which grew slowly over the next half year. Everything—even down to office supplies and telephones—remained scarce. But far more trying was the fact that he found himself beset by strong personalities. Bell's chief engineer Robert Stanley matched Williams point for point in stubbornness, determination, and intelligence. The two dueled often during the ten months leading up to the attempt to cross the sound barrier.

Stanley wanted to expedite the flight program at almost any cost in order to limit Bell's obligations. Williams, schooled by John Stack, insisted that no corners be cut, either in extent of instrumentation aboard the X-1s or in the full collection of flight data.

By and large, the Army Air Forces officials at Muroc and Dayton sided with Williams. Yet, he had additional headaches. The Army assigned a single-minded young pilot to fly through the sound barrier, one who resented Williams's caution and his insistence on gaining an understanding of the transonic region, rather than merely crossing it. Captain Charles E. "Chuck" Yeager, a World War II double ace and two-time winner of the Silver Star, possessed a great deal of native intelligence, if not formal learning. But Yeager felt patronized by Williams and his college-educated staff and resented the NACA and its time consuming methods. For the most part, however, the two men managed to coexist with one another.

On balance, their relationship assumed less importance than the actual flights of the X-1 aircraft. Yeager began the run-up to Mach 1 on August 29, 1947, when he flew to Mach 0.85. True to form, Williams chastised him for exceeding the planned speed of 0.80 and missing the data points. Nevertheless, by the fourth flight, in mid-September, Yeager pushed the aircraft to Mach 0.92. But these missions and those remaining on the way to Mach 1 proved to be treacherous and fraught with some hard lessons about the transonic region. At some moments the X-1's nose pitched up (Mach 0.87); at others it nosed down (Mach 0.90). Yeager felt mild buffeting at some speeds, severe at others. Most disturbing, elevator control declined sharply between the

YEAGER WATCHED AS THE NEEDLE ON THE ACCELEROMETER FROZE, AND THEN JUMPED TO THE RIGHT AND OUT OF VIEW, INDICATING THAT HIS AIRCRAFT HAD SURPASSED THE FABLED SONIC BARRIER.

Between September 1946 and October 1947, the small staff at the NACA's Muroc outpost (left), seen with the X-1 and the B-29 mothership behind them, prepared to challenge Mach 1. Bell Aircraft made two versions of the original X-1: the thin-winged Air Force model (X-1-1) and the thicker winged NACA variant (X-1-2), both flown over Rogers Dry Lake, 44 square miles (114 sq. km) of hard, flat surface, perfect for emergency landings by Muroc's aviators. Before such landings (or take-offs), the X-1 pilots tested the cramped interior of the aircraft, much as research pilot John Griffith does as he leans out of its hatch (opposite, top). Other preparations occurred in the hangars and workplaces that dotted the Muroc flight line. For instance, Jack Russell, Chief of the Rocket Shop, is seen here (below) conducting pressurization checks on the X-1's XLR-11 engine. At last, X-1-1 took wing over the Mojave, and after only eight envelope expansion flights during August, September, and early October, on October 14, 1947, Captain Chuck Yeager surpassed the sound barrier at Mach 1.06 (opposite, bottom).

first appearance of the transonic shock wave on the wings (Mach 0.88) and the passage of the wave over the aft end of the airplane (Mach 0.94). Some feared that this phenomenon posed such a danger that the project might need to be aborted, or at least postponed.

But an antidote presented itself. Before the Mach 1 flight, engineers at Muroc adjusted the plane's horizontal stabilizer in order for it to move freely, an innovation developed at Langley and adapted by Bell Aircraft in the design of the X-1. This step, taken in concert with his flight engineer Jack Ridley, enabled Yeager to retain longitudinal control through the treacherous transonic zone.

During the actual historic mission on October 14, 1947, Yeager penetrated this region with adequate handling, and regained full effectiveness at Mach 0.95 and higher. As he accelerated, he watched as the needle on the accelerometer froze, and then jumped to the right and out of view, indicating that his aircraft—nicknamed *Glamorous Glennis* in honor of his wife—had surpassed the fabled sonic barrier.

In the end, Yeager's achievement constituted an enormous success for the U.S. Air Force. But it signified an even greater victory for the NACA's John Stack, one of the first engineers to adopt the concept of transonic and supersonic flight, and the man who assembled the talent and wherewithal necessary to conquer Mach 1.

1. Thomas Wolfe, *Look Homeward, Angel*, quoted in James R. Hansen, *Engineer in Charge: A History of the Langley Aeronautical Laboratory, 1917–1958*, NASA Special Publication 4305, Washington, D.C. (NASA) 1987, p. 18.

> THE NACA NOT ONLY NEEDED TO SET UP HANGAR SPACE, OFFICES, AND HOUSING, BUT AT THE SAME TIME PREPARE FOR PERHAPS THE MOST IMPORTANT FLIGHTS SINCE THOSE OF THE WRIGHT BROTHERS.

⟵⋯ ↑ ⋯⟶ *NACA's Flight Test Unit at Muroc, California, soon became home to the eastern transplants who left the greenery of Langley to settle amid the parched terrain of the high desert. During good weather, they attended staff barbecues on a nearby ranch, like the one pictured here in 1949. During the cold of winter, when snow occasionally dusted the area, the female "computers" of Muroc assembled for an outdoor photo. These women performed the important task of reducing raw flight data etched on strips of film during the research flights, rendering it usable by the Muroc engineers. Muroc research pilot*

Joe Walker displays the playfulness and informality common to the western setting, leaping onto an X-1A aircraft like a cowboy mounting a horse. Right, Langley's John Stack, the man responsible for initiating the Muroc experiment and one of the first to delve into the mysteries of transonic flight.

2

Up and Out
From Mach 1 to Project Mercury

↗ George W. Lewis (right), the formidable and savvy first director of the NACA. Lewis took a keen interest in the projects and personnel of the Langley Flight Research Center, but confined most of his activities to fighting the bureaucratic battles for the agency in Washington, D.C., leaving the able Henry Reid to handle the details in Hampton, Virginia. Hugh L. Dryden (left) succeeded Lewis in September 1947 and ran the NACA with a firmer hand and a more fully articulated overall strategy.

Within weeks of the epochal flight of the Bell X-1, a cluster of institutional events occurred in the NACA and the Army Air Forces that collectively almost equaled the long-term importance of John Stack's historic orchestration of the Mach 1 achievement.

First, on September 18, 1947, Congress declared American air and ground forces to be independent of one another, establishing the United States Air Force. From its very inception, the Air Force oriented itself towards technical as well as bureaucratic independence. Informed in its infancy by the brilliant and flamboyant Hungarian-American scientist Theodore von Kármán, the Air Force opened a series of state-of-the-art in-house laboratories that rivaled those of the Army, the Navy, and even the NACA. (See more on von Kármán, below.)

These military developments might have raised doubts about the rôle, and indeed the survival, of civilian aeronautical research in postwar America, had it not been for a pivotal change in the NACA leadership. The venerable George W. Lewis, still only in his early sixties during World War II, had devoted himself heart and soul to the NACA during the conflict, taking no time off between Pearl Harbor and the cessation of fighting. Lewis and the Langley researchers found themselves almost overrun with war-related tasks, the main example being aerodynamic drag cleanup investigations on virtually the entire U.S. fleet of military aircraft. Exhausted from his labors, Lewis suffered two heart attacks in November 1945 which left him unable to fulfill the demands of his

job. When it became clear that he would not fully recover, Lewis's retirement became inevitable and he resigned from the NACA on September 1, 1947.

Lewis left an institution clearly on the rise. Indeed, by the time he stepped down as director, the NACA counted almost 2800 employees and had a budget of almost $31,000,000. Under Lewis's watchful eye it had developed into a national institution, with a chain of impressive laboratories that spread from coast to coast. World War II had given Lewis his big chance to build the agency. In order to assure the NACA's continued operation if the eastern seaboard underwent attack, Congress appropriated funds to build the Joseph S. Ames Aeronautical Laboratory in Moffett Field, California. It opened in 1940. Almost a mirror-image of Langley in mission and equipment, the Ames facility also served the function of demonstrating the NACA's supremacy over aeronautical research in the western United States, a region in which Theodore von Kármán's Guggenheim Aeronautical Laboratory had made some powerful inroads with local aircraft manufacturers.

Two years later, the NACA opened the Aircraft Engine Research Laboratory in Cleveland, Ohio, renamed the George W. Lewis Flight Propulsion Laboratory in 1949 and now known as the John H. Glenn Research Center. In its early years it specialized in turbine engine development, an area in which the military services felt the NACA were deficient during World War II. Finally, Lewis succeeded in opening the Auxiliary Flight Research Station (which was redesignated the Pilotless Aircraft Research Station in the year of its opening, 1945). Located in remote Wallops Island, Virginia, this facility—latterly called the Pilotless Aircraft Division—actually corrected another perceived lapse of the NACA during World War II: the capacity to conceive, fabricate, and test missiles and rockets.

The day after Lewis left the NACA, his successor walked into his office. His name: Hugh Latimer Dryden. No one seemed surprised by the choice. Lewis selected him, the NACA's Main Committee endorsed him, and its chairman, Jerome C. Hunsaker, welcomed him as an old friend.

Indeed, Dryden had been involved with the NACA almost as long as Lewis, although in a far different capacity. At the age of twenty-one, the young physicist established the National Bureau of Standards' aerodynamics section, a wind tunnel group first subsidized by the NACA in 1921. During the early and mid-1920s, he and Lyman J. Briggs, a colleague at the bureau, published several seminal NACA Technical Reports about a phenomenon known as compressibility, the characteristics of airflow in the transonic region.

Dryden rose steadily through the ranks of the Bureau of Standards, eventually becoming the chief physicist and later, associate director. Along the way, he established himself as one of the world's foremost scientists of flight, publishing papers that improved wind tunnel accuracy and verified experimentally the boundary layer theory expounded by Professor Ludwig Prandtl of Germany. During his climb to prominence, Dryden began a close and lifelong friendship with the Hungarian émigré Theodore von Kármán, an association that transformed Dryden's

Hugh L. Dryden

Hugh L. Dryden (born in 1898) started life modestly. The son of Samuel, an intelligent but unsuccessful schoolteacher who eventually became a streetcar conductor, and Zenovia, a quiet and upright woman, Hugh spent his early years in southern Maryland. Here, his father's ancestors had been farmers and tradesmen for two hundred years. But pressed by hard economic times, Samuel moved the family to Baltimore in 1908. In the big city, young Dryden not only flourished but showed signs of brilliance. He graduated from Baltimore City College (actually a high school) first in his class at the age of fourteen. He then entered Johns Hopkins University on a full scholarship, earned an undergraduate degree in mathematics, and under the tutelage of none other than Joseph Ames, received a doctorate in applied physics at the age of twenty. Ames then got him started at the National Bureau of Standards as an inspector of World War I munitions gauges. Dryden stayed at the Bureau for twenty-eight years, ultimately becoming its associate director. He served as director of the NACA from 1947 until its closure in 1958, after which time he became the first Deputy Administrator of NASA (under administrators T. Keith Glennan and then James Webb), in effect acting as the agency's senior technical adviser. Dryden, who died in 1965, received the National Medal of Science posthumously from President Lyndon B. Johnson.

One of the closest collaborations in twentieth-century American science, Hugh L. Dryden (far left) and Theodore von Kármán (third from left) also exercised a profound influence over international aeronautical research. The two are shown here in Europe at the end of World War II leading a mission of U.S. scientists on a survey of military aeronautics conducted on behalf of General of the Army Air Forces Henry ("Hap") Arnold. Their inquiries resulted in a milestone technology forecast known as Toward New Horizons, a report that dominated postwar American airpower doctrine and practice.

[Top left] *Posed here on Rogers Dry Lake, the Bell X-1E extended the flight research repertoire of the earlier X-1s. Bigger, heavier, and more powerful than the one that carried Chuck Yeager to fame in 1947, the X1-E also had a cockpit canopy and an ejection seat, both lacking in the others. Hugh Dryden pressed hard to get funding to transform the NACA's X-1 into the X-1E in his drive to pursue not just supersonic but also hypersonic flight.*

[Top right, left and above] *NACA High-Speed Flight Station pilot Joe Walker examines the X-1E in 1958, the last of the three years it flew. Prior to launch, technicians subjected the aircraft to an exhaustive checklist as they attached it to its mothership. By the 1950s, ground crews no longer had to lower the research aircraft into a pit, tow the bomber (in this case, a B-29) over it, and mount it in the bomb bay using belly straps. By then, hydraulic lifts enabled the X-1E to be elevated and the B-29 to be lowered for mating.*

↑ *Two more views of von Kármán, one earlier in his career, one later. In the first (above), he appears at Langley in 1926 in the center of the picture, legs apart, wearing a double-breasted coat. Max Munk stands near Kármán in the front row, wearing a hat; George Lewis is at the far right, also in a hat. In that year, the Hungarian-born Kármán toured the United States from coast to coast, eventually arriving at his future place of employment, the California Institute of Technology. Thirty-four years later, the aged Kármán is seen with Jet Propulsion Laboratory director William Pickering, on the left, and former student Frank Malina. Kármán founded JPL during the 1940s with the assistance of Malina and a few others.*

Theodore von Kármán

Born in Budapest, Hungary, in 1881, Theodore von Kármán first arrived in the United States in 1926 after a long period as director of the Aachen Aerodynamics Institute in Germany, which he transformed from a backwater school into one of the aeronautics powerhouses in Europe. A brilliant, dynamic, and popular teacher capable of reducing the most complex phenomena to recognizable examples from everyday life, Kármán attracted some of the finest professors and students, in part due to his reputation as a preeminent scholar, in part due to his charm and persuasive powers with patrons and industrialists. Like Max Munk, he studied with Ludwig Prandtl of Göttingen. Kármán's breakthroughs on boundary layer airflows and turbulence theory won him international acclaim. He began his career in the United States in 1930 as director of the Guggenheim Aeronautical Laboratory at the California Institute of Technology (GALCIT). He remained there until the late 1940s, when he returned to Europe as a founder of the NATO Advisory Group for Aeronautical Research and Development (AGARD), one of the many multinational scientific organizations he nurtured and participated in. He died in Aachen in 1963.

life. Inspired by Kármán's example as an organizer of international aeronautical research, Dryden joined the circle as well, and soon assumed the rôle of one of its most influential leaders.

Although close collaborators, the two men could not have been more different. In contrast to the warm and gregarious Kármán, friends and co-workers found Dryden to be approachable but modest and reserved. Equally distinct from the charismatic and chummy Lewis, Dryden nonetheless possessed his own soft-spoken charm. But most of all, his uncanny technical instincts commanded respect and adherents.

From his first days as George Lewis's successor, the forty-nine year old Dryden knew exactly the direction in which he wanted to guide the NACA. He decided that the agency needed to reorient itself toward travel at supersonic and hypersonic speeds, and perhaps sometime in the future, toward spaceflight. Associated with transonics since his youthful research on compressibility, and more recently the chairman of the NACA's subcommittee of high-speed aerodynamics, Dryden knew as well as anyone the technical aspects of conquering and surpassing the speed of sound. He also

knew the bureaucratic challenges firsthand, having sat on the original NACA Research Airplane Committee which conceived the Bell X-1 and the Douglas D-558.

Consequently, within a month of assuming leadership of the NACA, Dryden made two symbolic but telling gestures. He traveled the long distance to the isolated Muroc Army Air Field in California's Mojave Desert to observe personally the preparations for the historic supersonic flight set for October 1947. When he returned to Washington, he signified the value of the Muroc Flight Test Unit by revoking its temporary status and making it a permanent NACA station (like Wallops, under the stewardship of Langley). This decision gave Walt Williams, now director, a powerful advantage in his dealings with strong Air Force personalities like Chuck Yeager and formidable industry leaders like Bob Stanley of Bell Aircraft. It also enabled his group to expand in number and to undertake a host of projects beyond the scope of the X-1.

Dryden supported the high-speed initiatives pursued by Williams and his team through important collaborations and friendships that he forged during and after World War II. He realized that the success of the NACA, relatively small and anonymous among federal

Walter C. Williams

Walter C. Williams, born in 1919 in New Orleans, Louisiana, received a bachelor's degree in aeronautical engineering from Louisiana State University at the age of twenty. He then joined the Langley Memorial Aeronautical Laboratory's stability and control group where he made significant contributions to the design and improvement of many World War II aircraft. After the successful X-1 glide flights at Pinecastle, Florida, the Langley leadership asked Williams to lead the Muroc, California, Flight Test Unit. The forceful and energetic Williams not only supervised the X-1 and D-558 projects but later oversaw the X-4, X-5, XF-92A and B-47 projects, alongside the Century Series fighter research and the early X-15 development and flight tests. In summer 1959, NACA director Hugh Dryden asked Williams to assume an even more pivotal assignment as Director of Flight Operations for Project Mercury. Williams retired from NASA in 1982 as the agency's chief engineer, and died in Tarzana, California, in 1995.

agencies, depended on such cooperative ventures. Moreover, the institutions with which he joined forces not only knew him well but also knew the NACA well. Since its inception, Langley depended on the goodwill of the Navy and the Army not only for almost all of the experimental aircraft it flew but for the very ground on which it operated. In addition, some of its most successful projects originated with the military services.

Dryden expanded this tradition. He became well known in the Navy due to his successful leadership of the Bat air-to-surface weapon, the first self-correcting radar-guided missile that proved itself under fire. Indeed, Dryden and his team not only conceived, fabricated, and tested the Bat but at the end of the war actually witnessed its success against Japanese warships in the Pacific. He became equally well known in the Army and later the Air Force during his wartime collaborations with his friend von Kármán. Together, they led a distinguished group of scientists who contributed to *Toward New Horizons*, the famed postwar aeronautical forecast for General Hap Arnold. The two men subsequently created the USAF Scientific Advisory Board, the model for countless science panels that followed.

↑ The young and fiery Louisianan Walter C. Williams, picked by John Stack and Langley chief pilot Mel Gough, led a small group of engineers to the barren Muroc Army Air Field in September 1946. Here they conducted historic flight research on the Bell X-1 aircraft. Williams struggled not just with the technical problems; he also found it difficult to find adequate housing and office space for his growing staff.

↘ ⟶ During the late 1940s and 1950s, the NACA conducted extensive research into supersonic and hypersonic flight at the NACA High-Speed Flight Research Station on Edwards Air Force Base, California; the base's 132 members of staff are shown here assembled for a group photo in 1950. On Edwards, flight at high speed became commonplace, although the cost could be great. In May 1948, Howard Lilly died in the line of duty when his D-558 Skystreak aircraft crashed after take-off due to a failure in one of its turbine engines.

As director of the NACA, Dryden often turned his many military associations to advantage. For example, he mustered some of his many contacts in the armed forces to breathe new life into the X-1 project. It seems that by 1951, the engineers and pilots assigned to the NACA's X-1 (number 2) had nearly finished its research program, completing pressure distribution as well as lift and drag research. But rather than retire this warhorse of supersonic flight, Hugh Dryden, among others, thought of a new purpose for the NACA's X-1. He and his colleagues proposed replacing its relatively thick 10% to chord wings with radically thin 4% ones. Not only recent Langley wind tunnel results but also actual flight research data, comparing the NACA 10% to the USAF's 8% wings (on the X-1 number 1), predicted better transonic flight with the still thinner airfoils.

The leaders of the NACA sought funding for the project in a customary way: by appealing to the generals at Wright Field, who declined to help due to the austere budgets of the postwar period. Then Hugh Dryden got involved. Soon thereafter, officials at USAF headquarters expressed an interest in sponsoring the project, at which time the officers at Wright Field somehow found the

⌁ ⟶ The NACA staff on Edwards improved their offices in the mid-1950s when a new building arose on the northwest edge of Rogers Dry Lake. The groundbreaking occurred in January 1953, with Walt Williams turning the shovel as (left to right) Chief of Instrumentation Gerald Truszynski, Chief of Operations Joseph Vensel, Chief of Personnel Marion Kent, and a state official look on. In August 1954, most of the 250 employees—double the number of four years earlier—assembled in front of the just-completed administration complex for a photograph.

NATIONAL ADVISORY COMMITTEE FOR AERONAUTICS

N A C A

HIGH-SPEED FLIGHT STATION

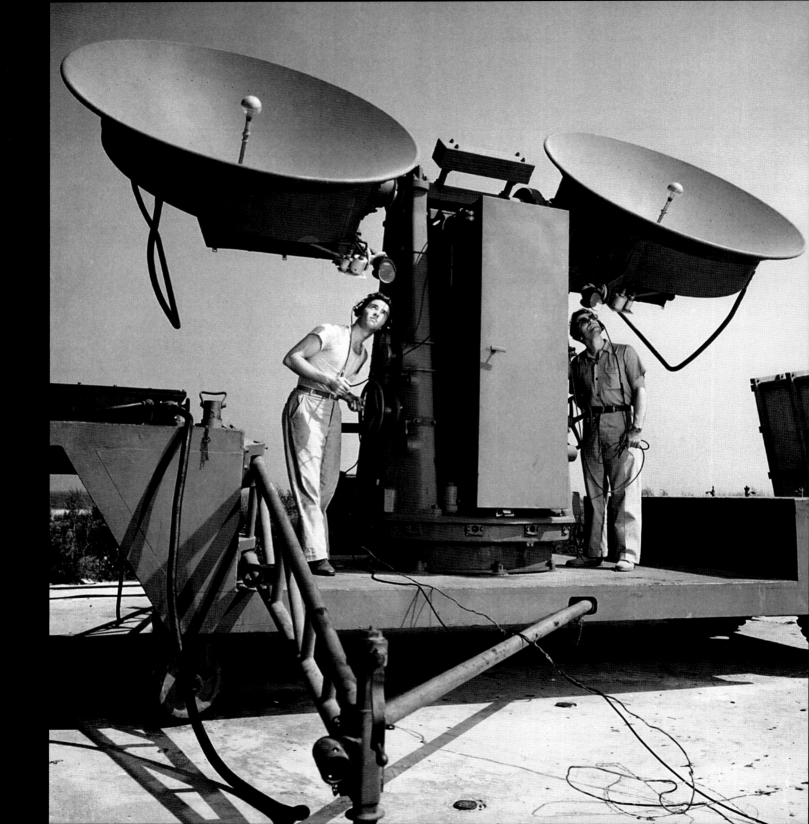

Like the flight research operation in California, the Wallops Island, Virginia, Pilotless Aircraft Research Division (a reporting unit of the Langley laboratory) undertook high-speed flight, but in unmanned vehicles. The first test firing (of a Tiamat missile) occurred there in June 1945 (right). Elaborate equipment soon followed, including Doppler radar recorders to track the trajectories of the rockets (far right). Outside, the radar dishes themselves (opposite)—produced by Sperry Gyroscope and known as the Model 10 Velocimeter— enabled tracking up to 5 miles (8 km) and captured data relating to speed and drag in flight. Despite the public impression at the time of the Sputnik launches that the NACA had been caught completely unprepared, nine months earlier scientists and engineers had successfully launched a five-stage model rocket from Wallops (top row).

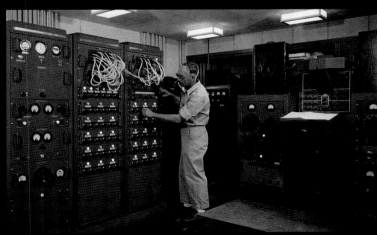

← ↗ *After the founding of NASA and the initiation of Project Mercury, Langley and Wallops engineers developed the short, barrel-shaped Little Joe launch vehicles and then flew them to test the Mercury capsule and recovery systems in flight. The Little Joe contained in its airframe a cluster of four Sergeant solid fuel rocket engines developed at the Jet Propulsion Laboratory in Pasadena, California. Illustrated, a Little Joe in its launcher and another lifting off.*

money to back the NACA request—a massive $900,000, drawn from no less a source than the Secretary of Defense's emergency funds. The resulting X-1E research aircraft eventually flew twenty-six times between 1955 and 1958, proving the value and the durability of the thin wings through the transonic range and up to Mach 2.2. It also produced preliminary data about the aerodynamic forces likely to be encountered by hypersonic vehicles.

Dryden pursued other, even more significant collaborations with the uniformed services in order to advance his high-speed agenda. After World War II, the American aviation industry, the armed forces, and the NACA all vied for government support of massive, high-speed and hypersonic wind tunnel construction. Tough negotiations continued for some time and appeared deadlocked. In the end, Hugh Dryden played a decisive role in finding a means to conciliate the parties. Rather than waste time bickering, he brokered a compromise, represented in the National Unitary Wind Tunnel Act of

1949. As a result of its terms, the NACA abandoned the hope of a Supersonic Research Center, instead ceding the facility to the Air Force (later embodied in the Arnold Engineering Development Center in Tullahoma, Tennessee). The nation's aircraft manufacturers also received an important concession, getting first use of whatever tunnels the NACA constructed as a result of the Wind Tunnel Act.

But Dryden negotiated well for the NACA. He won three National Unitary Wind Tunnels, costing an aggregate $136,000,000: a Mach 2 machine at Lewis, a Mach 3.5 at Ames, and a Mach 5 at Langley. As Dryden must have guessed, Boeing, Lockheed, and the other aircraft companies never claimed more than a fraction of the schedules of these great machines, leaving them essentially free for NACA use.

Dryden had good reason to accommodate his friends in the armed forces. During the early 1950s, the NACA initiated a battery of pivotal high-speed projects, some of

↑ ---→ Founded in the midst of World War II, the NACA Aeronautical Engine Research Laboratory (next to Cleveland, Ohio's Municipal Airport) undertook a broad range of research, from turbines to nuclear and hydrogen propulsion, for aircraft and for rockets. Named just after the war for George W. Lewis (renamed in 1999 the John H. Glenn Research Center at Lewis Field), the facility has employed both ground testing and flight research. Pictured right is the dedication of the lab in May 1943 (with Orville Wright, arms folded), and, above, the recording of flight data aboard a B-29 flying laboratory.

which required military sponsorship, others not. Part of the research occurred at the remote Pilotless Aircraft Research Station in Wallops Island, Virginia. Here engineers and technicians completed an astounding 3000 rocket and missile launches between 1945 and 1957, on average 250 per year with an 80% success rate. These vehicles flew as fast as Mach 12 and presented opportunities to test heat-resistant materials, dynamic stability, and new aerodynamic concepts. The engine laboratory at Lewis adapted itself to these new flight regimes, providing many of the powerplants and propellants for the Wallops Island researchers and testing such exotic fuels as hydrogen and nuclear materials on its new 20,000 pound (9072 kg) test stand. By 1957, about one-fourth of the Lewis staff was working on projects relating to rocket propulsion. Indeed, despite the slow rate of budget increases of the NACA during Hugh Dryden's tenure, he succeeded in reorienting the NACA toward more space-related activities by reallocating existing resources. For instance, during 1954 roughly 10% of the NACA budget involved space-related research; a mere two years later—by scaling back other projects—Dryden devoted no less than 25% of the NACA's resources to hypervelocity and space flight.

In tight fiscal straits such as those experienced by the NACA during the 1950s, the NACA director searched for ways to obtain military backing for projects that broadened the NACA's portfolio and gave it a credible claim to this new area of flight. Two projects enabled him to pursue this goal. The first won the NACA much acclaim for achieving one of the greatest milestones in high-speed flight. Roughly paralleling the development of the Bell X-1,

the Navy-sponsored Douglas D-558 aircraft offered the NACA the opportunity to establish further its credentials in supersonic flight. Even though the Army Air Forces and Bell Aircraft received extensive assistance from John Stack and the NACA during the X-1's development, the D-558 really represented the Langley design philosophy in full. Stack and his colleagues wanted an aircraft capable of staying aloft for long periods—not merely punching through the zone above and below the sound barrier—in order to amass large quantities of data from the transonic region. Hence, when the Navy and the NACA took possession of the first version of the D-558, known as the Skystreak, they at last had an aircraft that fulfilled Stack's wishes, although not perfectly. Sleek and graceful to behold, it flew beautifully below Mach 0.75. Unfortunately above 0.75 it became increasingly difficult to control, resulting in 1948 in a crash and the death of NACA research pilot Howard "Tick" Lilly.

Then, the following year, Douglas delivered D-558 number 2, called the Skyrocket. Swept-winged unlike its predecessor, the Skyrocket flew sometimes with a turbine engine, sometimes with rocket propulsion, and sometimes with a combination of both. But like the Skystreak, it proved cantankerous to fly, subject in its early career to violent pitch up. Between 1951 and 1953, one of the NACA's most able research pilots, A. Scott Crossfield, flight tested the Skyrocket in order to understand this dangerous phenomenon. These hazardous flights acquainted him with all of the aircraft's peculiarities, and armed with this knowledge, he and Walt Williams collaborated with the Navy to surpass the USAF in the next great speed milestone. Just two weeks before

↓ ↘ *Contrasting views of the Lewis laboratory staff. The cavernous drafting room—among the first structures built on the new site—soon became a hub of activity, where draftsmen translated engineering ideas into plans. Equally important work occurred at the evening "smokers" hosted by lab director Edward "Ray" Sharp, at which engineers and scientists mulled over research problems in a congenial atmosphere.*

⟵··· ⟵··· *The new Ames laboratory geared up for war quickly, hiring men (and many women) from California's San Francisco Bay area and beyond. Here, a male supervisor scrutinizes the work of women making slide rule calculations. In a far less confined setting, an F-86 fighter aircraft—the front-line U.S. fighter during the Korean War—undergoes testing in the mammoth 40 by 80 foot (12 by 24 m) full scale wind tunnel (for many years the largest in the world) built on the Ames campus.*

↑ *The Ames Aeronautical Laboratory, opened just before World War II in Mountain View, California, bore the name of one of the NACA's pivotal founders, Joseph Ames. Conceived to offer many of the same facilities as Langley (in order for the NACA to bolster its presence on the West Coast and to preserve research capacity should either side of the U.S. suffer attack), Ames soon became a sprawling plant, spread across a former dirigible facility at Moffett Naval Air Field.*

the fiftieth anniversary of the Wright brothers' first flight, on December 3, 1953, Crossfield and the Skyrocket (equipped with the Navy version of the XLR-11 rocket engine used on the X-1) dropped from the belly of a B-29 bomber flying over the barren Mojave landscape. Crossfield accelerated in level flight for forty-five seconds on all four rocket chambers until the powerplant started to misfire, telling the pilot to cut the speed. Crossfield looked down at the Machometer; to his momentary astonishment, it read 2.05—more than twice the speed of sound, a new milestone in aviation annals.

Shortly after the Skyrocket triumph, Dryden took steps to build on this success and take a full stride toward still faster flight. He and others in the NACA, in industry, and in the military services foresaw no major technical obstacles to the rapid increase in flight speeds, hastened in part by the powerful new rocket engines being fabricated for the nation's missiles. Not surprisingly, in this expectant atmosphere a number of ideas relevant to hypersonics manifested themselves almost at once. During 1950 and 1951, High-Speed Flight Research Station[1] engineers Hubert Drake and Robert Carmen proposed modifications of the Bell X-2 for Mach 3 flight, choosing this aircraft because of its K-Monel nickel alloy and stainless steel construction. About the same time, Dr. H. Julian Allen of the Ames Laboratory in Northern California discovered that blunt-nosed objects re-entering the Earth from space

[Bottom right]   *A Skystreak flies over the California desert. In contrast to the X-1, the Skystreak featured a cockpit canopy, but the extra height over the seat did little to add to the pilot's comfort. Because of the vehicle's slender profile, pilots complained of a cockpit so small that they could barely turn their heads. Taller men said that to read the instrument gauges they had to crane their neck downward, conforming their helmets to the contour of the canopy glass.*

[Opposite]   *In important respects, the D-558 Skyrocket had as much in common with the X-1 aircraft as with the Skystreak. Powered by a rocket engine similar to that of the X-1, the Skyrocket could take off from the belly of a bomber like the X-1. But the photograph of the Skyrocket on Rogers Dry Lake illustrates a sharp distinction from both the straight-winged X-1 and Skystreak: the Skyrocket had swept wings.*

[Above]   *Unlike the U.S. Air Force-inspired X-1 aircraft, the D-558 Skystreak and Skyrocket (D-558-1 and -2) represented the NACA's— specifically, John Stack's— research ambitions. Designed less to attain high speed than to probe the little known transonic region of flight and gather reliable data, these two very different vehicles flew collectively from 1946 to 1958. Illustrated, a D-558 model undergoing tests in the Langley 8 Foot High-Speed Wind Tunnel in 1947.*

[Top row and right]   *Three different views of the D-558-1 Skystreak aircraft. Unlike the X-1 series, the Skystreaks took off on their own power (rather than being lifted by a bomber and drop-launched), propelled by turbojet engines. Walt Williams's staff packed the Skystreaks with well over 600 pounds (272 kg) of instrumentation to record their behavior in flight. One is seen landing on Rogers Dry Lake, California, a second is parked on the ramp in front of one of the hangars at the Muroc Flight Test Unit, and a third serves as a prop for three of its crew members, also on Muroc.*

67

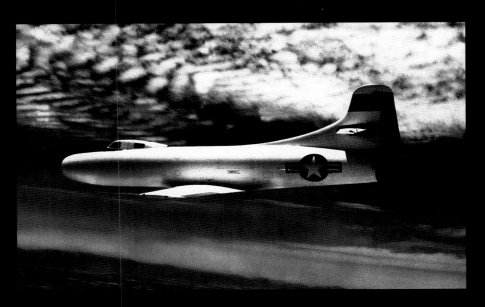

A. Scott Crossfield

Born in Berkeley, California, twenty-nine year old A. Scott Crossfield joined the NACA in June 1950 after enlisting in the Navy during World War II and flying F6F and F4U fighters. Between 1946 and 1950 he earned bachelor's and master's degrees in aeronautical engineering from the University of Washington. Crossfield remained with the NACA only five years but soon won a reputation as one of the agency's finest pilots, as well as a first-class engineer. In this short period, he flew some of the most pivotal experimental aircraft of his time, including the X-1, X-4, X-5, XF-92A, and the D-558 Skystreak and Skyrocket. Crossfield left the NACA in 1955 and joined North American Aviation, where he became a consultant on the famed X-15 project, making important contributions to its design. The first person to fly the X-15, he took its controls a further thirteen times, including the debut flight with the new and powerful XLR-99 rocket motor that eventually propelled it to a top speed of Mach 6.7. Crossfield left North American in 1967, after which he served as a technical consultant to the House Committee on Science and Technology. He received the prestigious Collier Trophy in 1961 from President John F. Kennedy.

encountered lower temperatures than those with a pointed shape, the result of a strong bow-shaped shock wave that deflected the heat. Meanwhile, Robert J. Woods of Bell Aircraft—designer of the X-1, X-2, and X-5 airplanes—encouraged NACA officials to consider the addition of hypersonic and even space research to the current projects list. The NACA Aerodynamics Committee responded to these random opportunities in spring 1952. It recommended the initiation of a hypersonic program to MIT's Dr. Jerome C. Hunsaker, the chairman of the NACA's Main and Executive Committees, who in turn instructed all NACA labs and stations to begin to focus on flight up to and beyond Mach 10.

Langley's engineers and scientists assumed a pivotal role in the unfolding of the hypersonics drama. Discussions there gave rise to the bedrock concept: a piloted, rocket-powered aircraft capable of reaching the limits of the Earth's atmosphere, and flying home by controlled glide to a runway landing. Engineers at Langley, Wallops Island, and the High-Speed Flight Research Station responded to this challenge in 1953 by transforming the Drake–Carmen X-2 proposal into a Mach 4.5 vehicle fitted with reaction controls and launched by two expendable solid rockets.

But NACA headquarters, almost certainly reflecting the wishes of Hugh Dryden, saw things differently. Instead of transforming an existing research airplane, it asked its

⟨---- ----⟩ At times, Skyrockets flew below the supersonic range; at other times, they achieved record speed. Outfitted with turbojet engines and Jet-Assisted Take-Off (JATO) rocket canisters they took off like conventional aircraft. But endowed with rocket power, they exceeded Mach 1. In fact, looking at a Skyrocket instrument panel like the one depicted here, NACA research pilot R. Scott Crossfield became the first to cross the second great high-speed frontier when he pushed a Skyrocket above Mach 2.

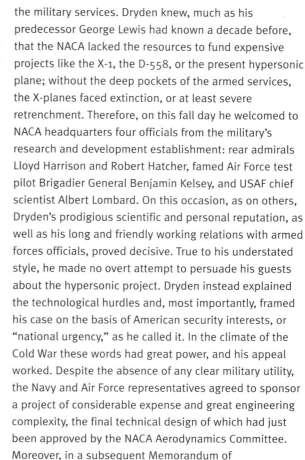

the military services. Dryden knew, much as his predecessor George Lewis had known a decade before, that the NACA lacked the resources to fund expensive projects like the X-1, the D-558, or the present hypersonic plane; without the deep pockets of the armed services, the X-planes faced extinction, or at least severe retrenchment. Therefore, on this fall day he welcomed to NACA headquarters four officials from the military's research and development establishment: rear admirals Lloyd Harrison and Robert Hatcher, famed Air Force test pilot Brigadier General Benjamin Kelsey, and USAF chief scientist Albert Lombard. On this occasion, as on others, Dryden's prodigious scientific and personal reputation, as well as his long and friendly working relations with armed forces officials, proved decisive. True to his understated style, he made no overt attempt to persuade his guests about the hypersonic project. Dryden instead explained the technological hurdles and, most importantly, framed his case on the basis of American security interests, or "national urgency," as he called it. In the climate of the Cold War these words had great power, and his appeal worked. Despite the absence of any clear military utility, the Navy and Air Force representatives agreed to sponsor a project of considerable expense and great engineering complexity, the final technical design of which had just been approved by the NACA Aerodynamics Committee. Moreover, in a subsequent Memorandum of Understanding signed by the NACA, the Navy, and the USAF at Christmastime 1954, the parties agreed to make Hugh Dryden the technical chair of the X-15 Research Airplane Committee. Finally, unlike the other X-planes agreements, in which the services conducted their own flight research programs, this one gave the NACA possession of all experimental vehicles and named the High-Speed Flight Station[2] as the site for all research flights. Designated Project 1226 by the Air Force, it became known more commonly as the X-15.

Despite the prestige inherent in being associated with a project of this magnitude, the winner of the competition to fabricate it—North American Aviation—actually declined to sign the contract. Company bosses attributed their decision to a backlog of orders, but the fact that the project consisted of only three prototypes (with no follow-on production contract) rendered it less than attractive from a profit-and-loss viewpoint. However, after some brokering by Hugh Dryden and General Howell Estes of Air Force Research and Development Command, North American president J.L. Atwood declared himself satisfied with a longer work period and extra start-up money from Department of Defense sources. Even so, despite an initial payout of over $40,000,000 for the three X-15s, North American's top leaders continued to fret about the project due to its low pay-off and fears that it might draw the

⤒ *The Bell X-2 represented a viable Mach 4.5 candidate (with modifications). In this photograph, it is shown dropping from the B-50 mothership in January 1955. Pilot Frank Everest flew it to Mach 1.40 on this flight, and attained a speed of Mach 2.87 later that year.*

field organizations to specify requirements for an all-new hypersonic airframe. Again, Langley stepped forward. A hypersonic panel was convened in Hampton, Virginia—led by the chief of the Compressibility Research Division, John V. Becker, and assisted by Langley rocket propulsion expert Maxime Faget—charged with conceiving the new airplane from a clean slate. By April 1954, they had arrived at the design of perhaps the most celebrated of all the X-planes: a vehicle of bold lines with a cruciform tail and wedge-shaped vertical fin, powered by three or four rocket motors, capable of Mach 7 speed and an altitude of several hundred thousand feet, launched from a mothership like the X-1s, and covered with a skin of the highly heat-resistant Inconel X chrome-nickel alloy. NACA headquarters accepted the concept and set the project in motion by assigning a distinct role to each NACA field office: flight planning to the High-Speed Flight Research Station, propulsion work to Lewis, aerodynamics studies to Ames, and hypersonic wind tunnel research to Langley.

Hugh Dryden had already played an important role in hypersonic flight, and with the unveiling of the Langley design he assumed the leadership of the project. Motivated by his lifelong intellectual interest in high-speed flight, as well as his ongoing campaign to focus the NACA's energies in that direction, he pursued the task vigorously and skillfully, but with characteristic reserve.

Dryden convened a meeting of the Research Airplane Committee in his office in October 1954. Just like the conference he had attended at Langley ten years earlier for the X-1, this one also included prominent members of

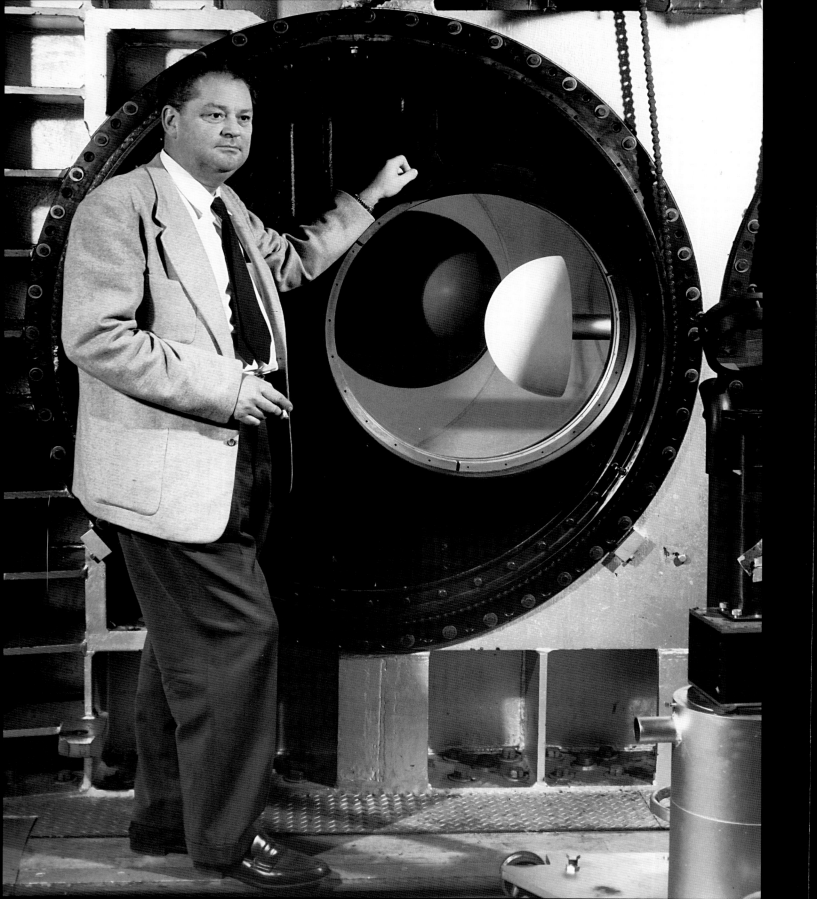

[Opposite] *Research scientist H. Julian Allen stands beside the observation window of the 8 by 7 foot (2.4 by 2 m) test section of the Ames National Unitary Wind Tunnel. Allen's insight about the superiority of the blunt body shape for re-entry vehicles revolutionized the design of ballistic missiles, as well as the Mercury, Gemini, and Apollo capsules, among others.*

[Top right] *During his long life, Jerome C. Hunsaker influenced American aeronautics in many ways. He founded the aeronautical engineering department at the Massachusetts Institute of Technology, became Chief of Design in the Naval Bureau of Aeronautics, and served for fifteen years as chairman of the NACA Main and Executive Committees, from 1941 to 1956.*

[Center] *Langley Compressibility Chief John V. Becker, standing next to the Langley 11 Inch Hypersonic Tunnel, played a crucial role in the conception of the X-15 hypersonic aircraft.*

Becker realized that the unusual confluence of technological possibility and political opportunity during the early 1950s made the X-15 possible, and he jumped at the chance.

[Bottom left] *In front of an X-15 are pictured three individuals who helped bring this and the other two hypersonic models to life. The program might not have existed without Hugh L. Dryden (center), who brought together the human and material wherewithal necessary for it to succeed. Flanking him are two men who guided the 199 X-15 flights: High-Speed Flight Station Director Walter Williams (right), and his successor Paul Bikle.*

[Bottom right] *Two years before the first flight of the X-15, Dryden visited the High-Speed Flight Station to present Chief Pilot Joe Walker (second from left) with the NACA Exceptional Service Medal. Also pictured are pilots Stan Butchart and Richard Payne, third and fourth from left.*

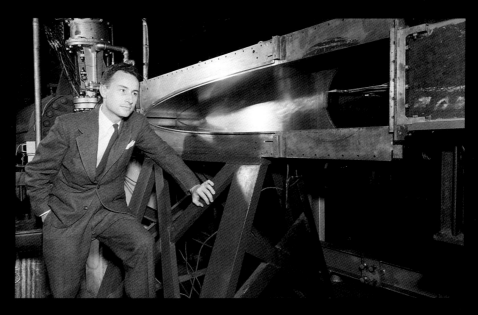

←···· *General Dwight David Eisenhower speaks at the Aircraft Engine Laboratory in Cleveland, Ohio, in 1946. As U.S. President, Eisenhower would support advanced research and development for aerial reconnaissance, ballistic missiles, and space-based overflight, but he backed civilian space activities less enthusiastically.*

EISENHOWER UNDERSTOOD THAT UNDER THE CONDITIONS OF INTENSE RIVALRY WITH THE U.S.S.R., AERONAUTICS REPRESENTED A POWERFUL INSTRUMENT OF NATIONAL POLICY.

company's best engineering talent away from more lucrative work. Luckily, Harrison A. "Stormy" Storms, the firm's able manager of research and development, took the program under his wing and headed a team of thirty-five designers, engineers, and technicians. As Storms's group prepared the first prototype, Reaction Motors won a contract to build the powerful XLR-99 rocket engine.

The NACA had a profound influence over the X-15 development. The specifications proposed by the Becker Committee dominated the project. Scott Crossfield, a man intimately involved with the X-planes, joined North American as a pilot and consultant in 1955 and made indispensable contributions to the X-15's design and fabrication. Walt Williams sent High-Speed Flight Station engineers and technicians to collaborate with their counterparts at North American. The Langley 9 inch (23 cm) blowdown wind tunnel conducted initial aerodynamics research on an X-15 model. The Lewis propulsion engineers in Cleveland made critical fuel recommendations for the XLR-99 engine to its builder.

Finally, at an unveiling at the North American plant in October 1958, the public caught its first glimpse of this black, sleek rocket plane. Long (49 feet/15 m), tall (14 feet/4.3 m at the vertical tail), heavy (31,275 pounds/14,186 kg launch weight, about twice that of the heaviest D-558), and powerful (57,000 pounds/ 25,855 kg of thrust from the XLR-99, once it worked), the X-15 breathed speed and modernity.

The very month in which technicians wheeled out the X-15 for photographers to shoot and the press to admire, a second revolution—entirely separate from the one led by Hugh Dryden—swept American flight. Based on decisions made at the highest levels of the U.S. government, these actions largely overshadowed the NACA's methodical buildup toward hypersonics and spaceflight. This national initiative had been some time in the making, but it occurred for the most part during the administration of President Dwight D. Eisenhower. The president, no stranger to Cold War realities, understood that under the conditions of intense rivalry with the U.S.S.R., aeronautics represented a powerful instrument of national policy, even as it continued to be a technology that shaped civilian life. As president, Eisenhower fostered this duality between civilian and military aeronautics and, to his credit, sustained it successfully.

Indeed, to a large degree, the former Army general hinged his Cold War strategy on controlling the world's airspace. Eisenhower inherited the U.S. presidency in the turbulent period of the Korean War. Like President Harry S. Truman before him, Eisenhower perceived the Soviets as aggressors and as the preeminent threat to American security. But Eisenhower also struck out in an entirely new direction from his predecessor by turning the struggle

↑ ↗ The Jet Propulsion Laboratory (JPL) engineers and scientists designed and fabricated the Corporal missile series under the sponsorship of the U.S. Army Ordnance Department, with the objective of fielding tactical guided missiles. Meanwhile, just after World War II, captured German scientists led by Dr. Wernher von Braun launched the famed V-2 rockets from White Sands (left). The two teams eventually collaborated, producing a two stage rocket (a modified V-2 first stage called the Bumper WAC, and a Corporal second stage) which flew to an altitude of 240 miles (386 km) in 1950 (right).

with Communism into a technological contest. He launched the development of Intercontinental Ballistic Missiles (ICBMs), advanced surveillance aircraft, and reconnaissance satellites, all of which inaugurated transformations in the evolution of flight.

In particular, as early as spring 1954, the president met with a team of scientists to discuss the use of space for the protection of the country. Both the memory of the sneak attack on Pearl Harbor thirteen years earlier and the U.S.S.R.'s preference for operating covertly convinced Eisenhower to begin research on aircraft and reconnaissance satellites capable of flying over Russian territory. Cloaked behind perhaps the tightest secrecy in the history of the nation, Lockheed Missile Systems received a contract in 1956 to fabricate a two stage rocket known as the Agena, as well as a satellite capable of photographing enemy territory with a wide array of standard, as well as invisible light, cameras. At the same time, Eisenhower accelerated missile and rocket testing. Two stage rocketry had actually been tested by the Army before the Eisenhower administration using captured German V-2 rockets mated to second stage WAC Corporal missiles, designed at the Jet Propulsion Laboratory in Pasadena, California. The Army followed the Corporal–V-2

combination with the Redstone rocket, flown thirty-six times from Cape Canaveral, Florida, beginning in 1953. (Its eventual mission involved the boosting of reconnaissance satellites into orbit.) The Naval Research Laboratory, meanwhile, conducted firings of its Viking sounding rockets, aboard which flew impressive photographic and high altitude research instruments. Finally, in 1955, the Air Force's Atlas ballistic missile underwent preliminary testing, and just four years later it became the first long-range weapon in the U.S. nuclear arsenal.

Even before the president undertook these transformative projects, civilian space activities (in addition to those being pursued by Hugh Dryden and the NACA) had also assumed a momentous role. At the start of the Eisenhower administration, the International Council of Scientific Unions declared the period between July 1957 and December 1958 to be the International Polar Year, dedicated to exploration of the remote parts of the Earth. Then, wishing to add research gathered from payload packages on sounding rockets, the organizers renamed the event the International Geophysical Year. The council further broadened the mandate in October 1954 by inviting countries to launch satellites to map the Earth's contours. The Soviet leadership accepted the

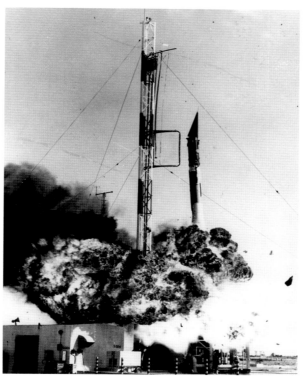
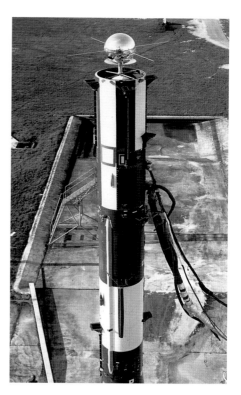

At the end of his second term as president, Dwight Eisenhower visited the Army Ballistic Missile Agency in Huntsville, Alabama, which he rededicated as the George C. Marshall Space Flight Center, in honor of his former comrade-in-arms. Here, Dr. von Braun shows the president a model of the first stage of the Saturn I rocket.

In pursuing the spaceflight challenge issued for the International Geophysical Year in 1957 and 1958, the U.S. Navy and Army vied to represent American interests. The Navy's Project Vanguard, launched from a Viking rocket, set the goal of measuring the shape of the Earth. But the Viking proved to be problematical. Illustrated is a Viking rocket undergoing a static test in September 1955, one exploding on the launch pad in December 1957, and a third on the launch pad in September 1959, awaiting installation of its nose cone.

challenge. President Eisenhower did as well, so long as the effort contributed to his doctrine of "freedom of space," a slogan that reflected a genuine American desire for open civilian exploration of the heavens. Meanwhile, with the president's blessing, U.S. agencies continued secretly to press with all their energy for the military option: to photograph enemy territory using reconnaissance aircraft and satellites.

At this point, a competition almost as fierce as that of the two superpowers took shape. The U.S. Army and Navy vied for the honor of achieving the International Geophysical Year objective. At first, the administration decided to back the Navy's Project Vanguard, powered by the Viking booster. Unlike the Army's Redstone, Viking had not been entangled in the ballistic missile program. But the two competitors advanced slowly under tightly corseted budgets, and neither made great strides.

Then, the great shock occurred. Soviet scientists and engineers stepped into the breach and won the International Geophysical Year competition when *Sputnik I* and *Sputnik II* (October and November 1957, respectively) circled the Earth, emitting powerful radio signals. With this sound echoing across the U.S. political landscape, the Americans responded with two crash programs. The Army's modified *Jupiter-C* ballistic missile flew the *Explorer 1* satellite into orbit on the last day of January 1958. The Navy's Viking sent *Vanguard 1* circling the earth the following St. Patrick's Day. Both achieved

more compelling missions than the initial Sputniks: *Explorer*, mapping radiation zones around the globe predicted by physicist James van Allen; *Vanguard*, finding the world to be slightly pear-shaped rather than round.

But nothing changed the hard fact. The Soviet Union, considered at the time to be technologically inferior to the United States, not only got into space first but threatened to open a lead. Many Americans reacted to this drama with dismay, disbelief, and anger. Many blamed President Eisenhower, who in reality had been assembling the ingredients of a powerful military space program that one day helped regain US hegemony in civil space as well. But the president and his government, bound by secrecy, found themselves unable to explain the plans that had been developing under wraps. Eisenhower had no option but to ride out the furor enveloping the country and his administration, which partly stemmed from a press eager to manufacture headlines. But the journalists had plenty of collaboration from members of Congress, hitherto indifferent or half-hearted about the almost microscopic NACA budget, but now steaming with rhetoric.

The nation faced a crisis. In order to avoid or at least minimize further Soviet triumphs, American political leaders joined a debate about the type and extent of a national space program. In due course, all eyes turned to that most retiring of federal agencies, the National Advisory Committee for Aeronautics, for answers. But first, Congress acted. In the month following the launch

The U.S. Army responded to Vanguard with a combination: a Jupiter-C intermediate-range intercontinental ballistic missile (developed by the von Braun team) and a satellite named Explorer 1, conceived at the Jet Propulsion Laboratory (JPL). Pictured, a Jupiter-C undergoing assembly at the Redstone Arsenal in Huntsville, Alabama (right); the mating of Explorer 1 to the Jupiter-C (top); and the combined stack, awaiting launch on January 31, 1958 (above). JPL also contributed the two stages above the Jupiter-C, consisting of a cluster of Baby Sergeant (or Juno) rockets.

TOP UNIT-RE-ENTRY

[Opposite, below and bottom right] *America achieves sustained spaceflight. On January 31, 1958, the Jupiter-C/Juno/Explorer 1 combination rose on the launch pad at Cape Canaveral under the scrutiny of a team of mission controllers in a nearby blockhouse. Once* Explorer 1 *achieved orbit, JPL's Dr. William Pickering (left), Professor James Van Allen, and Dr. Wernher von Braun lifted a model of the satellite in triumph for the press.*

[Right and far right] *Before either* Vanguard *or* Explorer 1 *reached the heavens, the U.S.S.R. succeeded in sending* Sputnik I *into Earth orbit on October 4, 1957. The satellite is shown here on a rigging truck in the assembly shop as technicians prepare it for flight. More shocking still, the Soviets followed it with a second launch the following month.* Sputnik II *(seen here covered by a payload shroud on the launch pad at Tyuratam, U.S.S.R.) carried the dog Layka into space, but she died after four days aloft.*

↖ During his tenure as director of the NACA (1947 to 1958), Hugh L. Dryden worked expeditiously to transform the agency from one devoted to aeronautics to one committed to hypersonics and even spaceflight. Lacking a national consensus to achieve this end, he accomplished it within the constraints of politics and the NACA budgets. Even though he failed to win the job as NASA leader, thanks to his foresight NASA came into being with a fully articulated space program, ready to be enacted as soon as Congress appropriated funds.

↑ The last meeting of the NACA occurred on August 21, 1958, attended by the first Administrator of the National Aeronautics and Space Administration, T. Keith Glennan. Sworn in just two days earlier, Glennan chats with two members of the NACA Main Committee, Charles McCarthy of Vought Aircraft, and Preston Bassett, member of the NACA Aerodynamics Committee. At Glennan's insistence, Hugh Dryden agreed to serve as his deputy.

of *Sputnik I*, Senate majority leader Lyndon B. Johnson chaired a series of hearings about American defense and space programs in the Armed Services Subcommittee on Preparedness. As a result of Johnson's vigorous investigation, covered by a highly enthusiastic press, a gathering consensus concluded that a coherent national space program needed to be inaugurated.

Unfortunately, President Eisenhower lacked similar enthusiasm for such an initiative. He felt, with some justification, that he had already safeguarded the nation's defense by beginning the reconnaissance and the ballistic missile programs, and regarded a direct competition with the Soviets as a mere sporting event. But afraid that Johnson and the Democrats might seize the initiative, the president attempted to get ahead of events. He renamed his existing and somewhat anonymous science panel the President's Science Advisory Committee and appointed MIT's distinguished president, Professor James Killian, to be its chair. He directed Killian to do nothing less than draft a comprehensive civil space policy for the United States.

During the winter months of 1957 and 1958, Congress and the White House vied for definition and mastery of this new and unfamiliar entity. The president had a clear objective. Openly concerned about the growing influence of the American military–industrial alliance — or complex,

⤏ An aerial view of the Jet Propulsion Laboratory, located in Pasadena, California. Surrounded by the San Gabriel Mountains, JPL began as an outgrowth of rocket research led by Theodore von Kármán at nearby Caltech during the 1930s and 1940s. The Arroyo Seco canyon, where JPL began with a few Works Progress Administration shacks, eventually developed into the huge campus shown in this photograph.

as he called it—that had been forged in the cauldron of Cold War politics, Eisenhower refused to consider the creation of a space agency within the Defense Department. Some in Congress looked to the Atomic Energy Commission as an existing structure capable of managing such big technology projects. But the president and his staff had other plans. At last, in February 1958, James Killian issued the findings of his panel: facing a continued challenge from the Soviet Union, the nation required a space agency, one organized around the nucleus of the NACA. Hugh Dryden and others in the NACA felt vindicated.

In a short time, however, a different story was to emerge. The legislation defining the new organization originated with Killian's staff, but the NACA and Bureau of the Budget officials abetted the process. The proposal submitted to Congress provided for a civilian National Aeronautics and Space Agency, headed by an Administrator chosen by the U.S. president and confirmed by the Senate. Nothing in the wording gave particular alarm to Hugh Dryden or to James Doolittle (NACA chairman since Jerome Hunsaker's departure in 1957), even though it abandoned the committee structure by which the NACA had operated since 1915. Eisenhower approved the plan and although he and Lyndon Johnson wrangled over the powers of an advisory panel to guide overall space policy, the president signed the act in July 1958. The National Aeronautics and Space Administration (NASA) came into being on October 1 of that year, submerging—but in a sense preserving—the NACA in the new organization.

Fortunately or not, Congress still had a great deal to say about the birth of this new institution. Nevertheless, Hugh Dryden felt confident of his chances to transcend the metamorphosis from the NACA to NASA and retain his leadership rôle. He even asked his friend and fellow science administrator H. Guyford Stever (later President Richard Nixon's science adviser) to consider serving as his deputy. But most members of the Senate never considered Dryden seriously, despite his eminence as a scientist and his successful tenure as NACA director, during which time the NACA made steady strides toward hypersonic flight, and even spaceflight.

For one, Dryden seemed temperamentally unsuited to the position. In a time when the nation was experiencing the shock of taking second place to the Soviets in a head-to-head technology race, Dryden simply seemed too reserved and self-effacing for the job of NASA Administrator. Politicians instead looked for someone who would make bold predictions and present a bright future. For his part, Dryden disdained the role of cheerleader and made no secret about it, saying on more than one occasion before congressional committees that space

←--- ↑ *Known as the Dolly Madison House because the wife of President James Madison lived there for many years, this historic building at 1520 H Street, NW, Washington, D.C., served as the first NASA Headquarters. T. Keith Glennan—shown here in his official portrait—served as Administrator between 1958 and 1961. A graduate of Yale University in electrical engineering, Glennan worked in the motion picture industry in the development of talking pictures, served as director of the U.S. Navy Underwater Sound Laboratories during World War II, and subsequently became president of Case Western Reserve University in Cleveland, Ohio. He returned to Case after leaving NASA.*

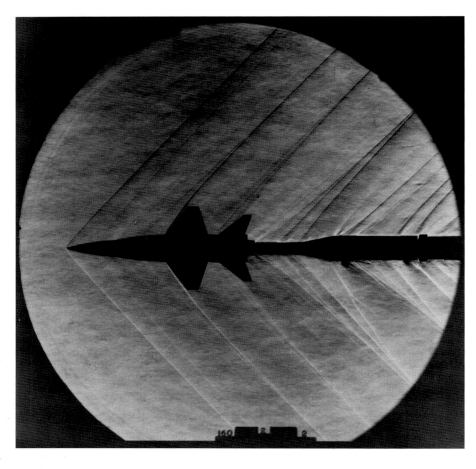

scientific circles, as well as his technical reputation, Hugh Dryden enjoyed almost universal recognition and had contacts everywhere. For these reasons, he exercised a profound influence over the Mercury, Gemini, and Apollo programs.

Of course, the original NACA laboratories and stations—Langley, Lewis, Ames, Wallops, and the Flight Research Station[3]—played transformative rôles in the development of the new agency. But many more constituent elements joined them. Just after the signing of the National Aeronautics and Space Act by the president, other federal entities began to be absorbed into the existing NACA structure. First, six weeks after NASA came into existence, Project Vanguard and its 150 employees transferred from the Navy to NASA, eventually forming the nucleus of the Goddard Space Flight Center in Greenbelt, Maryland. Next, in December 1958, the Army-sponsored Jet Propulsion Laboratory in Pasadena, California (despite its name, devoted to rocket research), fell into the NASA orbit. Glennan also asked for and received the transfer of the Army Ballistic Missile Agency in Huntsville, Alabama, in July 1960, but not without loud protests from the service, which was determined to maintain a credible rocket program. The Ballistic Missile Agency offered two great prizes: the German rocket scientist Dr. Wernher von Braun and his staff; and the space vehicle under design by his team, the Saturn rocket, capable of 1.5 million pounds (680,000 kg) of thrust in its initial stage alone. Finally, the Air Force relinquished development of the mighty F-1 rocket engine to NASA.

Meanwhile, during the year of political and bureaucratic tribulations following Sputnik, the X-15 program represented one bright spot in a U.S. space program struggling to get its footing. The fabrication of the three airframes at North American proceeded relatively well. The propulsion system proved to be less certain. Despite intense pressures to respond to recent Soviet space successes, North American's partner in the hypersonic project, Reaction Motors, conceded in 1958 that the revolutionary XLR-99 powerplant needed further work before being mated to the X-15. In the interim period, the parties agreed to rely on a pair of XLR-11 engines, like those that propelled the X-1 aircraft. As it turned out, the twin XLR-11s needed to be used during the entire first year of the X-15's flight program as XLR-99 engineers at Reaction encountered serious developmental problems.

By late 1958, the X-15 program had become freighted with expectations. High Speed Flight Station director Walt Williams expressed this sentiment when he told a group at North American that he expected the hypersonic aircraft to answer many of the open questions about the eventual role of human beings in space exploration.

The initial X-15 flight—the first in a series by North

exploration existed solely to expand scientific horizons for the benefit of the country as a whole. Moreover, many of the legislators seemed to hold him accountable for Sputnik. To some extent, Hugh Dryden became the public figure blamed for the Soviets' space success, even though Congress failed year after year to provide budgets that might have equipped the NACA to pursue a national effort.

In light of these political liabilities, President Eisenhower unwillingly abandoned Dryden—who had devoted his career at the NACA toward supersonic, hypersonic, and spaceflight—and instead nominated the president of Case Western Reserve University, T. Keith Glennan. Glennan agreed to become NASA Administrator, but on the condition that Dryden serve as his deputy and technical adviser. Although hurt at being denied the top post, Dryden stayed on. This decision proved to be invaluable to Glennan's and NASA's early successes. Indeed, Dryden's experience repaid NASA and Glennan many times over. As the author of the NACA's space policy, Dryden knew its innermost details, as well as the personalities involved. He was also well acquainted with the corridors of Congress and had developed a finely tuned instinct for bureaucratic choices. Finally, due to his ubiquitous presence in American and international

American—occurred on Edwards Air Force Base in June 1959, about eight months after the demise of the NACA. Like many to follow, it encountered problems. Fortunately, the pilot in the cockpit came well prepared. Scott Crossfield—the conqueror of Mach 2 who had been hired by North American at the inception of the X-15 program—profited not only from his personal knowledge of the high speed experience in two earlier rocket planes (the X-1 and D-558) but also from his talents as an engineer. Indeed, Crossfield served as a midwife to the X-15, contributing extensively to its design and getting to know its peculiarities like no one else.

His experience proved to be life-saving. During the maiden (unpowered) flight on June 8, 1959, as he waited in X-15 number 1 (itself lashed to the underbelly of the B-52 mothership), he noticed that the pitch damper failed to operate. The flight rules left no choice but to abort the mission, but Crossfield had the authority to overrule them and did so, thinking the immense lakebed and the uncomplicated flight path offered some safety. However, as so many first flights showed, the unexpected reigned supreme. As he approached for a landing and attempted to reduce the steep glide path, longitudinal (nose to tail) oscillations started, a condition ever more dangerous as his air speed decreased. Crossfield managed to land at the bottom of one of the oscillations without injury, but the impact caused extensive damage to the landing gear and resulted in six months of repair. Project engineers combated the problem by increasing the rate on the horizontal stabilizer actuators, and it never occurred again.

Crossfield subsequently piloted the "Black Bull" (as

Wernher von Braun

Wernher von Braun began life in a world of privilege. His father, Magnus von Braun, a Prussian aristocrat, became a banker during the 1920s. Wernher was born in Wirsitz, Prussia, in March 1912. Falling under the spell of Romanian spaceflight visionary Hermann Oberth during his teens, he decided to devote his career to rocketry. Von Braun graduated from the Berlin Technical University and earned a doctorate in rocket combustion in 1934. He became the technical director of the German rocket laboratory at Peenemuende, on the Baltic Sea, which in 1942 achieved the first successful firing of the V-2 ballistic missile, later used to attack targets in the United Kingdom. Peenemuende under von Braun made extensive use of forced labor. At the end of the war, Project Paperclip—a secret American program to recruit German scientists—brought von Braun and many of his associates to the United States, first to Fort Bliss at the White Sands Testing Range, New Mexico, where he and his team demonstrated the V-2 rocket to his Army sponsors. After being transferred in 1950 to the Army's Huntsville, Alabama, Redstone Arsenal, he directed the Army Ballistic Missile Agency's Development Operations Division. In 1960, von Braun became the first director of NASA's Marshall Space Flight Center. Ten years later he accepted the post of Associate Administrator for Plans at NASA headquarters. He died in 1976.

⟨····⟩ Dr. Wernher von Braun and his group at the Redstone Arsenal (and later Marshall Space Flight Center) developed many of the launch vehicles used during the early U.S. space program. A charming and likeable figure, he nonetheless remained controversial because of his past collaboration with the Nazi party in Germany. Still, he made decisive contributions to the American space program. Von Braun is shown in his office at Marshall in 1964 and in a T-38 aircraft during a visit to the NASA Flight Research Center, Edwards Air Force Base, the previous year.

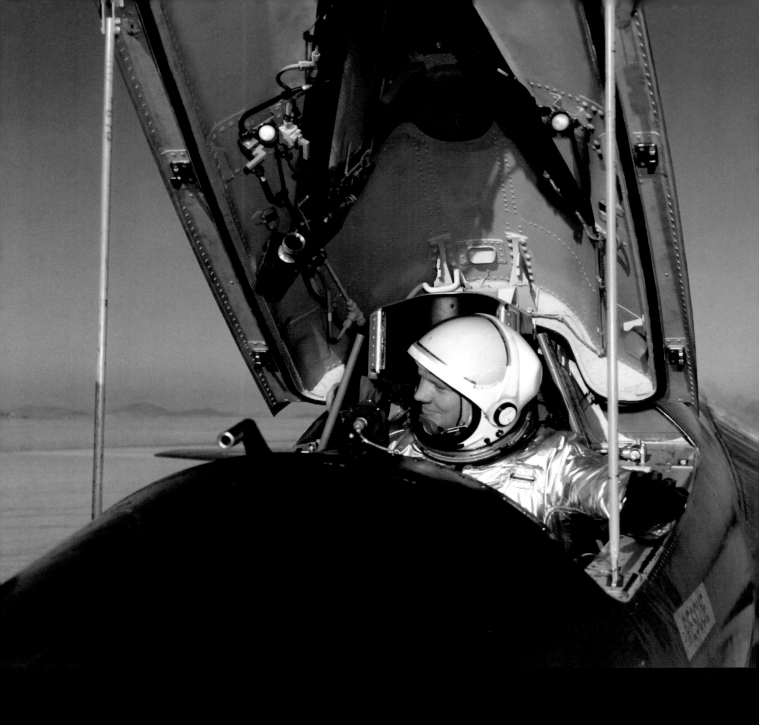

NASA research pilot Milt Thompson called the powerful but unpredictable machine) thirteen more times, three of which assumed pivotal importance: the first powered flight (using the XLR-11 engines) in September 1959; the first flight in the new and untested XLR-99 in November 1960; and in the same month, the first trial of the XLR-99 to re-start and throttle in flight. Although other X-15 pilots flew much faster and much higher, Scott Crossfield assumed perhaps the greatest risks as he took all the initial steps that made possible the vehicle's subsequent research.

Ultimately the X-15 was highly successful, and it represents for many a harbinger of true spaceflight. One of the most important successes was in terms of its psychological value. During the long three and a half years between the success of *Sputnik I* and the first rocket launch of an astronaut into space, the X-15 program represented America's only human-centered space flights. Indeed, before the first suborbital mission (May 1961), the highly photogenic X-15 aircraft flew thirty-six times over the course of nearly two years, and magazines and newspapers covered it widely. Thus, the X-15 paid dividends in a way Hugh Dryden never intended: it represented a welcome introduction to spaceflight for the American people, a gentle initiation into the hair-raising, rocket-powered ascents in tiny metal capsules that followed.

Defenders of the X-15 pointed out that it gave the taxpayer much more than an interlude before the U.S. space program found its footing. It actually served its purpose in several ways. To begin with, it inducted a group of eight Americans into the astronaut corps—that

is, those who traveled at least 50 statute miles (264,000 feet/80,467 m) above the Earth.

Although he did not fly to this height, the nation's most celebrated astronaut trained in the rudiments of spaceflight as an X-15 pilot. Neil Armstrong, then a NASA research pilot, flew the X-15 seven times between November 1960 and July 1962, achieving hypersonic speed (Mach 5.74) and very high altitude (over 207,000 feet/ 63,094 m). Indeed, on his fourth flight, he experienced the vagaries of travel far above the Earth when he fell victim to a slight "bounce" out of the atmosphere as he descended back into it. This caused him to lose momentarily the capacity to turn, forcing him off his flight path and resulting in a perilous landing in which he touched down about 10 miles (16 km) away from the appointed spot.

Eventually, the X-15 flight envelope expanded to Mach 6.7 (4520 miles/7274 km per hour) and to an altitude of 354,200 feet (107,960 m). During the physiologically demanding conditions experienced by the X-15 pilots in flights such as these, aeromedical researchers wired them with monitoring devices to check their vital signs, measure vertigo, and record other factors that collectively constituted a library of human responses to spaceflight. This data informed those planning for the bigger space missions to come.

The X-15 scientists and engineers also discovered the potential stresses to spacecraft associated with high-speed travel through the atmosphere. During the envelope expansion phase of the program, they carefully instrumented the aircraft and sought to determine the characteristics of hypersonic aerodynamics and the

↑ *A photograph of the hypersonic X-15 number 2 just after launch in the early 1960s. Although subject to intense heating due to the effects of high speed aerodynamics, the aircraft here shows patches of frost in the middle, resulting from the liquid oxygen in the propulsion system. Liquid nitrogen cooled the payload bay, cockpit, nose, and windshields.*

effects of thermodynamic heating on the vehicle's skin and structure. They uncovered several unexpected phenomena. Despite the overall accuracy of the wind tunnel tests of the aircraft's aerodynamics, the aft end of the vehicle experienced 15% more drag in flight than predicted. Moreover, doubling the speed of the X-15 vastly increased the temperature effects. Researchers calculated the heating load at Mach 6 to be eight times that at Mach 3. In addition, some parts of the airframe—the front and the lower portions especially—had greater susceptibility to damage from high temperature than others. The plane's raised cockpit also generated turbulence and heating problems, resulting in damage to the X-15's windshield on two occasions. Other factors such as handling qualities, stability augmentation systems, and reaction controls rendered the X-15 one of history's most important aerospace testbeds. Overall, aerodynamicists learned a decisive lesson from the X-15: safe flight at hypersonic speeds demanded that even small design features of marginal importance for subsonic flight be given the closest attention in vehicles exceeding Mach 5. Ultimately, the Space Shuttle orbiter would become the greatest beneficiary of these and many other observations gained from the data gathered during the X-15's 199 flights.

In its later flights the X-15 contributed to the U.S. space program as a platform for hundreds of scientific experiments. For instance, this flying laboratory collected micrometeorites to assess their possible impact on space

Robert R. Gilruth

Born in 1913, Robert R. Gilruth joined the Langley Memorial Aeronautical Laboratory in 1937 after graduating from the University of Minnesota with a B.S. and M.S. in aeronautical engineering. He quickly became involved in the NACA's influential flying qualities experiments, soon took the lead in them, and by his early thirties was acting as chief of Langley's Flight Research Section. During World War II he pioneered transonic research using dive tests on highly instrumented P-51 Mustangs. At the end of the war he formed a team of Langley engineers to study experimental rocketry, an initiative that resulted in the NACA test range at Wallops Island, Virginia. During the 1950s, Gilruth became the assistant director of Langley. After leading the Space Task Group during the early stages of Project Mercury, he assumed the role of director of the newly formed Manned Spacecraft Center in Houston, Texas (later the Lyndon B. Johnson Space Center), and held the job until 1972. After a period of service at NASA headquarters, he retired from the agency in 1983. Robert Gilruth died in the year 2000.

⟵⋯⟶ *Instructed by NACA director Hugh Dryden to begin planning for Project Mercury months before NASA came into being, Robert Gilruth assembled a team of engineers at Langley (known as the Space Task Group) to devise the basic architecture of America's first program to place human beings in orbit. By the time of NASA's founding on October 1, 1958, the project had been substantially conceived, enabling the new agency to waste no time in responding to the Soviet challenge. In August 1959, Gilruth and his deputy, Charles Donlan (far left), inspected a model of the Mercury capsule.*

vehicles. It carried an infrared scanning radiometer that recorded the Earth's radiation from 70,000 to 100,000 feet (around 21,000 to 30,500 m). Finally, four of the X-15's long duration forays into space involved tests of a zero-gravity heat exchanger designed to cool future spacecraft.

Just as engineers at the North American plant prepared the first X-15 for its rollout, events in Washington, D.C., opened a second front in the U.S. human spaceflight campaign. Once again, Hugh Dryden played a guiding role. During the early summer of 1958—still some time before NASA came into being—Dryden called Langley engineer Robert Gilruth to NACA headquarters and presented him with a mandate to map an initial American program of sustained spaceflight with human passengers. Gilruth assembled a panel of ten known as the Space Task Group, drawn from Langley and from the Lewis propulsion laboratory in Cleveland. During the following three months they conjured the main ingredients of Project Mercury.

In a briefing to Congress on August 1, 1958, Gilruth recommended an expeditious and uncomplicated approach that paralleled the proposals of another distinguished Langley researcher, Maxime Faget, first aired at the final NACA conference on high-speed aerodynamics six months before the birth of NASA. For the early suborbital missions, Gilruth's blueprint incorporated the short-range Redstone ballistic missile, developed by Wernher von Braun and his team at the Army Ballistic Missile Agency. For the later orbital flights, the Space Task Group chose the Atlas ICBM as the boost rocket. They conceived of a small capsule, mounted atop the Atlas, to lift the human payload. After a few

revolutions around the Earth, the spacecraft relied on a retrorocket to brake its momentum and drop it back into the atmosphere. The capsule was protected by a heat shield as it re-entered, and its final descent depended on the opening of a parachute, a splashdown in the ocean, and the pickup of crew and spacecraft by the U.S. Navy.

These spaceflight concepts became embodied in Project Mercury (named for the winged Roman mythological figure who served as the messenger of the gods). Led by Gilruth, Mercury foretold the more centralized authority that was to emerge gradually in the new agency. Gilruth reported not through his chain of command at Langley, but worked directly for another hand-picked appointee of Hugh Dryden, former Lewis engineer Abe Silverstein, the director of all space projects at NASA headquarters.

Despite the new order in Washington, D.C., the continued leadership of Dryden, Gilruth, and Silverstein in Project Mercury represented the NACA's last and most important gift to the nation, even as it passed out of existence and NASA rose in its place.

1. The Muroc Flight Test Unit—a tenant on Muroc Air Force Base, California (renamed Edwards Air Force Base in January 1950)—became known as the NACA High-Speed Flight Research Station in November 1949.

2. Formerly the High-Speed Flight Research Station, redesignated in July 1954.

3. The NACA High-Speed Flight Station (formerly the High-Speed Flight Research Station) became the NASA Flight Research Center in September 1959.

←···· Gilruth and his associates decided to launch the initial suborbital flights of Mercury with the most dependable rocket in the inventory, the U.S. Army Redstone. Developed by the von Braun group at Huntsville, this liquid propelled ballistic missile earned the nickname "Old Reliable." It is shown being prepared on the launch pad and at its first lift-off, on August 20, 1953.

····→ For the orbital flights of Mercury, Gilruth selected another ballistic missile, the Convair Company's Atlas ICBM. Modified as the Mercury-Atlas rocket, it underwent test launches to qualify its systems for spaceflight. In 1962 it stood on Pad 14 at Cape Canaveral, prepared for an orbital flight.

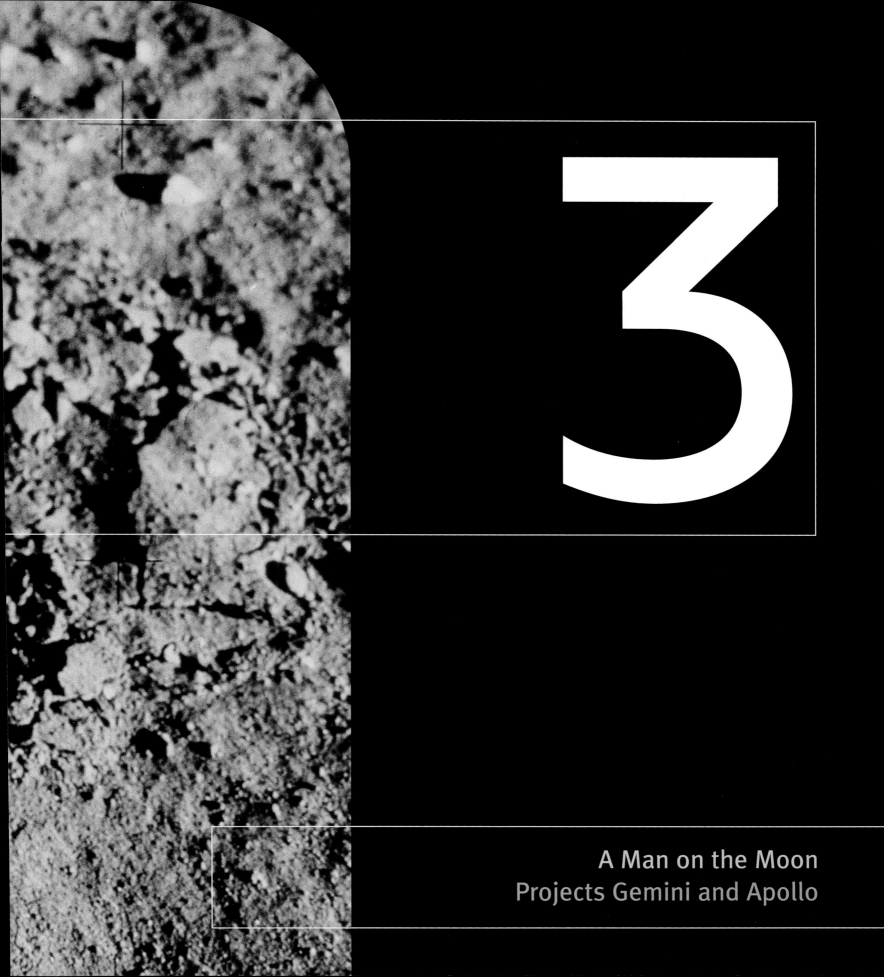

3

A Man on the Moon
Projects Gemini and Apollo

[Top left] *Fourteen individuals, who were later known throughout the world, posed for photographs after becoming astronauts. Astronaut Group 1, named in April 1959 to fly Project Mercury, included (seated, left to right) L. Gordon Cooper, Virgil I. Grissom, M. Scott Carpenter, Walter M. Schirra, John H. Glenn, Alan B. Shepard, and Donald K. Slayton. Named nearly three and a half years later, Astronaut Group 2 consisted of (standing, left to right) Edward H. White, James A. McDivitt, John W. Young, Elliott M. See, Charles Conrad, Frank Borman, Neil A. Armstrong, Thomas P. Stafford, and James A. Lovell.*

[Center left] *After three days of survival training in the Nevada Desert, fourteen astronauts stop in January 1964 to commemorate the moment. In the front row, left to right, are William Anders, Walter Cunningham, Roger Chaffee, Richard Gordon, and Michael Collins. Behind them (left to right) are Clifton Williams, Eugene Cernan, David Scott, Donn Eisele, Russell Schweikart, Buzz Aldrin, Alan Bean, Charles Bassett, and Theodore Freeman.*

[Below] *The Mercury Seven looking at a model of their future spaceship, just days after being selected in 1959. Seated (left to right) are Virgil (Gus) Grissom, Scott Carpenter, Donald "Deke" Slayton, and Gordon Cooper; standing (left to right) are Alan Shepard, Walter Schirra, and John Glenn. The Space Task Group at Langley (especially deputy STG Director Charles Donlan and research pilot Robert Champine) made the selections, using graduation from military test pilot school as a decisive prerequisite. The astronauts also needed to be younger than forty, be in excellent physical condition, be under 5 feet 11 inches (1.8 m) tall, hold a bachelor's degree in engineering, possess 1500 hours' flying time, and be qualified jet pilots.*

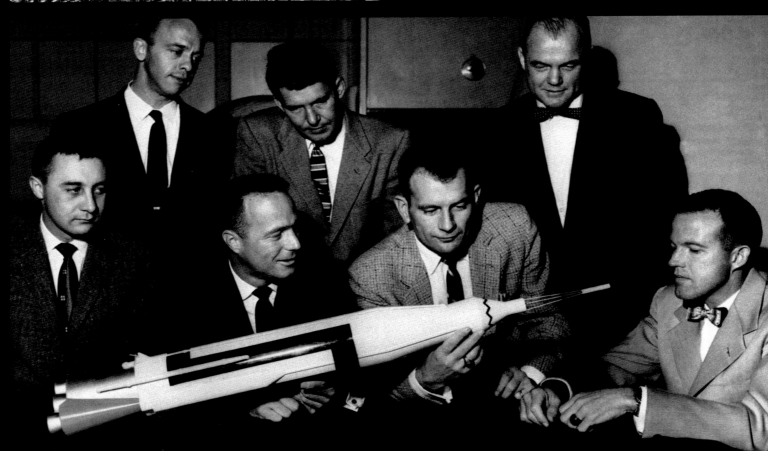

The legacy of Hugh Dryden and the NACA paid dividends to NASA almost from the day Congress established the new agency on October 1, 1958. Because so much planning and research had already been achieved at Langley, Ames, Lewis, and the High-Speed Flight Station, NASA opened its doors ready to compete with the U.S.S.R.

Almost immediately after taking office, Administrator T. Keith Glennan made a fundamental decision about America's first human spaceflight project. After hearing Gilruth and his Space Task Group describe their proposal, Glennan sent them back to Langley to assemble the core technologies required for Project Mercury. At least one of these requisites, the project's capsule, had already been conceived by Langley's Maxime Faget and several collaborators. They imagined a blunt body vehicle only 11 feet (3.6 m) long, just 6 feet (1.8 m) wide at its broad base, and having the capacity to sustain a single human being for up to twenty-four hours. Inside the cramped interior, the astronaut—strapped into a tightly fitted seat—would fly in orbit with his back towards the Earth.

McDonnell Douglas Aircraft won the prime contract for Mercury in 1959, representing a vast management departure from NACA tradition. Typically, the NACA's researchers acted as their own prime contractors on projects, calling upon companies for subsystems or components as needed, but more often than not fashioning these items in their own machine shops. Of course, during the mid-1950s the NACA operated on frugal annual budgets of about $100,000,000, enforcing the habit of thrift on its leaders. After Sputnik, however, time, not money, became the essential factor. In order to achieve objectives quickly, the new agency had no choice but to rely on outsiders to design and fabricate machines and equipment according to NASA specifications.

The following year, the project team assembled the main ingredients required for Mercury. Technicians integrated boosters with spacecraft. Ground stations started to open around the world in order to provide continuous tracking. A mission control center came into being.

Beginning in 1959, a rigorous astronaut selection process took place under Gilruth's supervision. In the end, seven military pilots—Marine aviator John H. Glenn; Navy fliers Alan B. Shepard, M. Scott Carpenter, and Walter M. Schirra; and Air Force officers Virgil I. "Gus" Grissom,

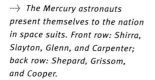
The Mercury astronauts present themselves to the nation in space suits. Front row: Shirra, Slayton, Glenn, and Carpenter; back row: Shepard, Grissom, and Cooper.

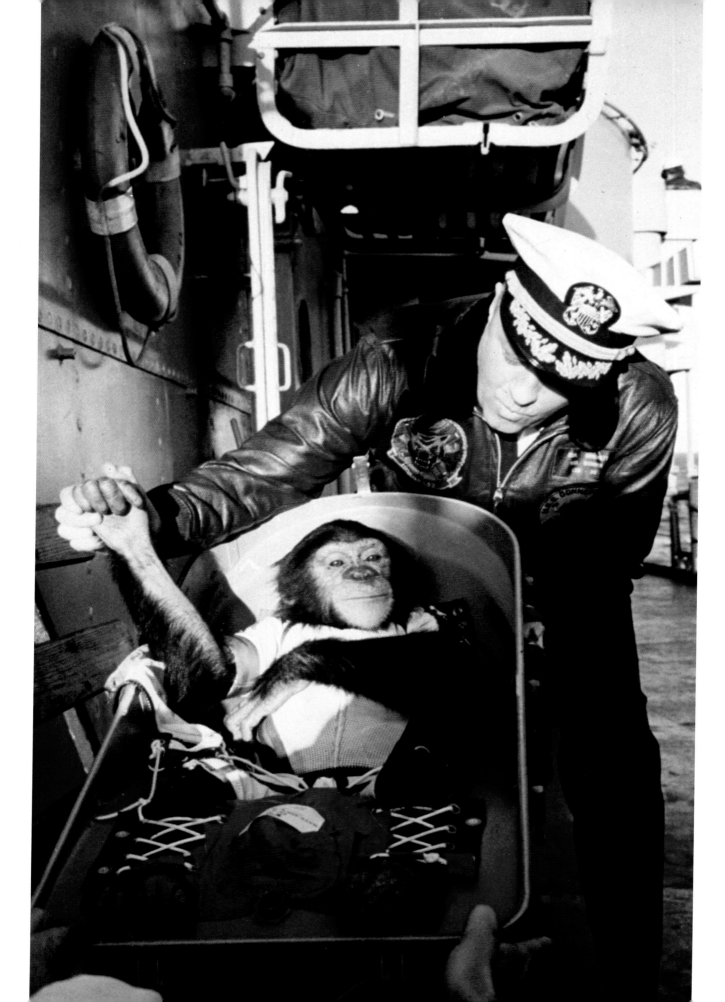

⟵ Before astronauts came chimpanzees. In order to test the physiological consequences of spaceflight without risking human life, NASA launched a chimp known as Ham aboard a Redstone rocket in January 1961. Ham greeted the commander of the recovery ship that plucked him from the Atlantic with a handshake.

↑ ↗ Converting a missile into a launch vehicle capable of carrying human cargo required extensive modification and testing. Here, on the Redstone Test Stand at Marshall Space Flight Center, a Mercury capsule and escape system undergo installation prior to a test firing. Important though the Mercury–Redstone combination proved to be for U.S. spaceflight, in just a few short years the first generation of launch vehicles seemed insignificant next to the third (a Redstone versus a Saturn I).

L. Gordon Cooper, and Donald K. "Deke" Slayton—appeared before a frenzied press in April 1959. The test firings of Mercury hardware began later that year; in August two Rhesus monkeys were flown on a cluster of small Little Joe rockets. Then the bigger Redstone launch vehicles, designed by the von Braun researchers at Huntsville, lifted astronaut dummies and chimpanzees into space. One, named Ham, traveled for nearly seventeen minutes and survived both the trip and the recovery.

For obvious reasons, the flight of human subjects proved more problematical. After many technical obstacles and postponements, on May 6, 1961, Alan Shepard rode the Mercury–Redstone combination in a successful suborbital foray into space. The second attempt, also suborbital and also fraught with reversals and delays, launched Virgil "Gus" Grissom on his mission on July 21, 1961, but nearly resulted in his death when the hatch of the *Liberty Bell 7* capsule blew off prematurely during the recovery at sea.

The remaining flights—all orbital—aroused much concern initially because of the launch system. The Atlas missiles on which they lifted off had been fabricated by engineers at Convair using less traditional methods than those employed by the Huntsville team on the Redstones.

Rather than achieving structural strength through heavy metal bracing and a thick outer shell, the Convair group built a thin-skinned booster strengthened by internal pressurization. In the final analysis, the Mercury–Atlas pairing performed superbly, beginning in November 1961 with the launch of Enos the chimp, who flew twice around the Earth before being recovered safely.

At last, John Glenn became the first American to attempt to circumnavigate the world from space. *Friendship 7* left Cape Canaveral aboard the Atlas on February 20, 1962, but encountered two frightening events. First, Glenn needed to take the controls from a malfunctioning autopilot on two of the three orbits. Then, during re-entry, a faulty switch indicated a loose heat shield. Partly because he surmounted these difficulties, Glenn returned to a degree of adulation unseen since Charles Lindbergh's epic transatlantic flight almost four decades earlier. Mercury evolved rapidly during the succeeding missions. By the final voyage, in May 1963, Gordon Cooper made twenty-two revolutions around the Earth over a period of twenty-two hours. If the X-15 program constituted a leap out of water, as Hugh Dryden called it, Mercury represented America's first taste of sustained space travel, not to mention a mine of precious

[Left] *Robert Gilruth and his Space Task Group decided to try two suborbital flights before attempting a more difficult and dangerous orbital one. In this photograph, John Glenn, Alan Shepard, and Gus Grissom pose for a publicity shot in the space suits, with the Mercury–Redstone stack behind them. The Redstone rocket carried only Shepard and Grissom.*

[Bottom row] *Selected for the first Mercury mission, Alan Shepard walks from the elevator atop the Redstone rocket to the cleanroom prior to squeezing into the narrow confines of the capsule, called Freedom 7. During his brief foray into space on May 5, 1961, a 16 mm movie camera captured his movements and expressions.*

[Opposite, top] *Shepard, his fellow astronauts, and the space agency itself failed to anticipate the outpouring of public interest in all the early space missions. Capitalizing on this fervor, on the day following Shepard's flight President John F. Kennedy presented him with the NASA Distinguished Service Award on the White House lawn. Shepard's wife (in hat), his mother, and the six other Mercury astronauts joined in the celebration.*

[Opposite, bottom row] *Gus Grissom, seen here being strapped into his flying gear by space suit specialist Joe Schmidt, made the second suborbital flight on July 21, 1961, after delays due to bad weather. Beforehand, assisted by astronaut John Glenn, Grissom angled himself into the Liberty Bell 7 capsule. He stayed aloft for sixteen minutes during a safe flight.*

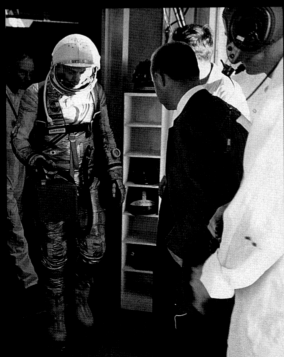

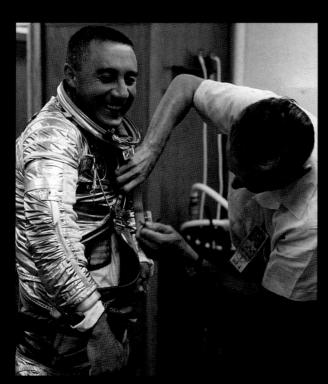

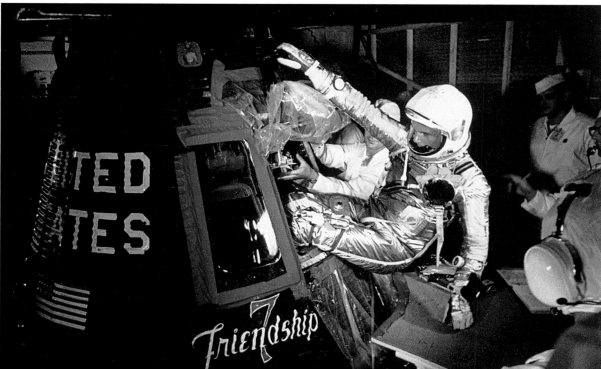

Aerial view of Complex 14 at Cape Canaveral, Florida, the launch pad of the orbital Mercury flights. The Atlas missile—the launch vehicle of all Mercury flights apart from the two suborbital missions—needed to be integrated with the Mercury equipment and tested thoroughly before being approved for human spaceflight.

Astronaut John H. Glenn, seen in his space suit at Cape Canaveral and entering Friendship 7 prior to launch on February 20, 1962. In a mission not lacking in technical problems, Glenn flew around the world three times. His flight nonetheless constituted a giant stride for the U.S. space program, in one blow advancing the duration of American spaceflight from fifteen minutes to almost five hours. Glenn's Friendship 7 capsule fell into the Atlantic near Grand Turk in the Bahamas, and twenty-one minutes later Navy destroyer Noah pulled him out of the sea.

data about human physiology in weightlessness and practical experience about the myriad engineering hurdles common to such missions.

Momentous as Project Mercury may have been, even more epochal events overshadowed it. A combustible mixture of presidential politics, Cold War rivalries, and technological opportunity conspired to imbue spaceflight with a gravity no one had predicted. The drama started in November 1960 with the election of President John F. Kennedy over Richard M. Nixon by the narrowest popular margin (fewer than 119,000 votes). Because of such a close vote, Kennedy seemed destined to govern cautiously. But instead, the new president—the youngest man (at forty-three years old) ever to be elected to the office—showed as much daring in his public life as his detractors claim he did in private. Kennedy ran for office, and won, on a theme of Cold War militancy, claiming that the United States needed to redress a large "missile gap" with the Soviets that exposed the nation to attack, or at least to nuclear blackmail. Although the charge proved to be chimerical, once in office the Kennedy administration needed to prove its Cold War credentials.

But rather than take on the Soviets head-to-head, the president—advised by the Joint Chiefs of Staff and the Central Intelligence Agency (CIA)—instead chose Cuba, one of the U.S.S.R.'s surrogates. Schooled and armed by the CIA, opponents of Fidel Castro mounted an invasion of the island in mid-April 1961. The attackers struck at the Bay of Pigs, but suffered a rout at the hands of Castro's forces. The resulting furor humiliated the president and his administration and occasioned celebration in Moscow and sympathetic capitals. Unfortunately, it came on the

James E. Webb

The son of a county school superintendent from Oxford, North Carolina, James E. Webb (1906–1992) showed a talent for organization and extraordinary energy even as a youngster. He studied at the University of North Carolina at Chapel Hill, and took a degree in education. After enlistment in the Marine Aviation Reserve, where he learned to fly, Webb went to Washington, D.C., and worked as an aide on Capitol Hill. When the position of Director of the Bureau of the Budget opened in 1945, President Truman appointed Webb on the strong recommendation of some influential friends. After four years there, he became undersecretary of state under the formidable Dean Acheson. Webb left Washington during the early 1950s, but returned in 1960 to re-enter public service. Despite his own misgivings about his fitness for the NASA position, President Kennedy and Vice President Johnson persuaded him to accept it. He did, but insisted that the indispensable Dryden remain as Deputy Administrator. Until Daniel Goldin during the 1990s, James Webb enjoyed the longest tenure of any NASA Administrator, serving from 1961 to 1968. His reputation suffered as a result of the January 1967 launch pad fire of *Apollo 1*, taking the lives of astronauts Grissom, Chaffee, and White. Still, a purposeful man who liked to lead, James Webb, who resigned under pressure from President Johnson before the ultimate success of *Apollo 11*, nonetheless succeeded in fulfilling the daring mission initiated by President Kennedy.

←···· *A mere fifteen months after John Glenn's triumphant mission, astronaut L. Gordon Cooper made not one but twenty-two revolutions around the Earth in the final Mercury flight. Cooper is shown in a pressure suit, holding his helmet. Cooper died in Ventura, California, in October 2004.*

····→ *James E. Webb was selected by President Kennedy to be the second Administrator of NASA in April 1961, succeeding T. Keith Glennan. Not technically schooled, Webb nevertheless proved to be a skillful and tireless champion of NASA in the counsels of government and perhaps more responsible than anyone but Kennedy himself for landing Americans on the Moon.*

heels of an equally momentous event. Less than a week before the Bay of Pigs catastrophe—on April 12, 1961—cosmonaut Yuri Gagarin, aboard *Vostock I*, became the first human being to orbit the Earth. The American response the following month only added to the sense of despair. Alan Shepard's suborbital flight aboard *Freedom 7* lasted a mere fifteen minutes.

Thus, President Kennedy, elected to be tough on the Soviets, found himself enmeshed in two punishing Cold War defeats just three months into his term. Then the president remembered a proposal made by the Administrator of NASA, James E. Webb. The shrewd and politically savvy Webb had held top positions at the Department of State and the Bureau of the Budget, and although not technically trained brought boldness and vision to the new agency. In March 1961, before Cuba or Gagarin, he asked the president to increase the NASA budget and launch a drive to land Americans on the Moon by the close of the 1960s. Not especially interested in or attuned to technology, and uncomfortable with the costs, Kennedy listened but declined. But in April, the idea sprang back to life in the wake of the dual blows to American leadership and pride. Kennedy reasoned that a bold declaration to reach for the Moon might return the political momentum to the U.S., although he still worried about the magnitude of expense. Nonetheless, he asked Vice President Johnson—as senator, one of the architects of NASA—to investigate the prospects for a lunar initiative.

After only a few weeks of investigations (accompanied by his own inimitable lobbying in Congress), Johnson told the president that the lunar landing and return might be achieved, but at a staggering cost (estimated by Hugh

Dryden at about $33,000,000,000, and by 1967 at the earliest). James Webb voiced support for a more firm and conservative deadline: the end of the decade. Kennedy accepted the challenge and transmitted it to Congress on May 25, 1961, in a famous address entitled "Urgent National Needs." Delivered just three weeks after America's first flight in space (and that one merely suborbital), the president's speech nonetheless offered a ringing declaration of national purpose (which substantially underestimated the project's true fiscal burdens):

Let it be clear—and this is a judgment which the Members of Congress must finally make—let it be clear that I am asking the Congress and the country to accept a firm commitment to a new course of action—a course which will last for many years and carry very heavy costs: 531 million dollars in fiscal '62—and estimated seven to nine billion dollars additional over the next five years. If we are to go only half way, or reduce our sights in the face of difficulty, in my judgment it would be better not to go at all.[1]

As Congress considered the president's request (which they did approve in the end), James Webb and his engineers at NASA realized that even if the Moon program received full funding, the agency still lacked the technological know-how to achieve Kennedy's objective. It became apparent that Mercury did not adequately prepare NASA for the great leap forward, and that before attempting the lunar surface some essential features of the endeavor needed to be tested. Many of the phenomena might be probed on the ground, but some demanded flight outside the atmosphere.

Much needed to be discovered. For one, Project

↓ *Once John F. Kennedy had committed his presidency and the nation to human spaceflight, he showed increasing interest in the technologies and personalities in the program. In September 1962, he heard a briefing from Army Major Rocco L. Petrone at Cape Canaveral about the Air Force's test annex there. Petrone, a tough, no-nonsense figure, later became NASA's Director of Launch Operations during Project Apollo.*

[Top left] *An understandably reflective Cosmonaut Yuri Gagarin looks out the window of a bus transporting him to the launch pad on April 12, 1961. On that date, the twenty-seven year old Gagarin flew spacecraft* Vostock I *once around the world in 108 minutes, becoming the first human being to fly in space. Gagarin later served as deputy director of the Soviet Cosmonaut Training Center. Gagarin died in an aircraft training accident in March 1968. His success in* Vostock I *helped prompt President Kennedy to initiate Project Apollo.*

[Top right and center] *President Kennedy examines a model of the Apollo command module in 1962 and receives an explanation about its role from Robert Gilruth, director of the Manned Spacecraft Center in Houston, Texas. Vice President Lyndon B. Johnson—a seasoned senatorial veteran of space policy and one of the architects of NASA—looks on at the far right. In August of that year, the president recognized Gilruth's contributions with a presentation on the White House lawn of the Distinguished Federal Service Medal. James Webb stands between Gilruth and Kennedy.*

[Right] *In October 1962, nine months after NASA authorized Project Gemini as a follow-on to Mercury and a preparation for Apollo, astronauts Elliott See and Ed White (right) greet and drink coffee with nurse Dee O'Hara as they prepare for later missions.*

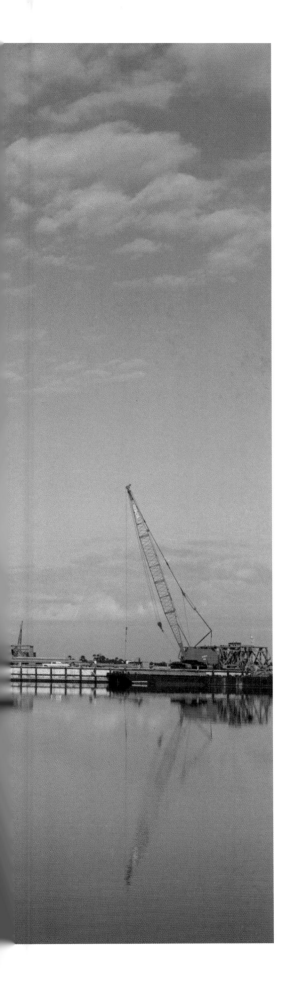

←···· *In the midst of projects Mercury and Gemini, construction of the Vehicle Assembly Building began at Kennedy Space Center (named for the late president and formerly the NASA Launch Operations Center at Cape Canaveral). An enormous, thirty-six story structure, the assembly building contained enough volume for technicians to integrate the parts of the Apollo–Saturn stack, as well as future launch vehicles.*

↘ *President Kennedy makes the case for the dangers and cost of spaceflight at an address at Rice University in September 1962, attended by about 35,000 people in the football stadium. Just a month before the Cuban missile crisis gripped the world, Kennedy declared the necessity of entering space in order to preserve it from international conflict. He spoke of the importance of the lunar adventure as a challenge to the nation, pursued, "not because it is easy, but because it is hard . . ."*

Mercury had not generated sufficient physiological data about prolonged exposure by humans to the space environment. Nor had it demonstrated in flight all of the design work necessary for a Moon launch. Moreover, no experience had yet been gained in the mechanics and safety of space walks, essential if human beings were to experience the cosmos first hand. Finally, there remained the untried and complicated procedure of rendezvous and docking of spacecraft, an inescapable ingredient of lunar travel.

Robert Gilruth and his Space Task Group at Langley took these problems in hand and conceived of Project Gemini, named for the mythological Greek twins Castor and Pollux, guardians of mariners. The Gemini spaceship embodied this duality: a modified and enlarged Mercury capsule capable of carrying two astronauts for long-duration flight. NASA authorized Gemini in January 1962. Once again, the U.S. civil space program turned to off-the-shelf military equipment to boost its hardware out of the atmosphere. Gilruth's team persuaded the U.S. Air Force to make available both the new *Titan II* missiles as Gemini's launch vehicles, and the Atlas–second-stage Agena combination as a target to practice rendezvous and docking. The Agena, equipped with a restartable engine, offered the advantage of being either a passive or an active partner to the Gemini capsule.

With the reality of Gemini and the looming lunar flights, NASA initiated a number of institutional adaptations, in part a response to the mammoth tasks ahead, and in part due to James Webb's desire for systematic and orderly procedures. To begin with,

[Opposite, top left and far left] *The first three American space programs devoted to human flight all landed in the ocean, a problematical method due to the size and vagaries of the sea, as well as the possibility of accidental drowning (as so nearly happened with Gus Grissom). During most of the decade in which the U.S. pursued early spaceflight, a different method of return to Earth received theoretical and practical attention by NASA. Two researchers at Ames Research Center—H. Julian Allen (left) and Alfred J. Eggers (far left)—contributed greatly to this new concept. Allen conceived the idea of blunt bodied vehicles as a method of reducing the heat of re-entry; Eggers designed an aircraft, the contours of which paralleled Allen's theory.*

[Opposite, center left] *Further translating H. Julian Allen's concept into a flying vehicle, NASA Flight Research Center engineer R. Dale Reed designed a lifting body, so named because of its high lift-to-drag ratio. Reed believed these aircraft had the potential to become the future of spacecraft, capable of returning to Earth by piloted runway landings. In this illustration, Reed holds in his hands a model of the first lifting body, known as the M2-F1, the real version of which is parked behind him.*

[Opposite, bottom far left] *Dale Reed's success converting his M2-F1 concept into a real (but unpowered) aircraft owed a great deal to the Flight Research Center's director, Paul F. Bikle. Bikle not only found the money to get Reed's wingless contraption*

fabricated, but in the end permitted one of his best pilots to fly it. It ultimately flew often over the barren landscape of Rogers Dry Lake, as did an entire family of lifting bodies developed jointly by NASA and the Air Force during the years that followed.*

[Opposite, bottom left] *NASA X-15 pilot Milt Thompson also backed Reed's lifting body idea. Indeed, Thompson persuaded a reluctant Bikle to let him test Reed's machine in flight. Before these experiments, Thompson (right), Gus Grissom (left), and Neil Armstrong tested another land-based re-entry concept at the Flight Research Center, the Paresev, a piloted glider considered (and then rejected) for Project Gemini.*

[Bottom right] *Soon, Milt Thompson (far left in this photo) had company in flight testing the gumdrop-shaped M2-F1. Not only did the hero of the X-1 (now Colonel) Chuck Yeager (seated in the cockpit) take the controls, but NASA research pilots Don Mallick and Bruce Petersen (to the right of Yeager) flew it, as did six others.*

[Top right and bottom far right] *Succeeding the M2-F1, the M2-F2 and the HL-10 transformed the lifting body concept from a curiosity to a real prospect. Here, the M2-F2 is shown undergoing wind tunnel tests at NASA Ames in 1965. One year later to the day, the HL-10 made its first flight at Edwards Air Force Base.*

[Top row] *As the lifting body research began to prove its value in California, the drama of Project Gemini unfolded at Cape Kennedy. During the Gemini 4 mission in June 1965 astronaut Edward White (flying in tandem with James McDivitt) achieved the first U.S. extravehicular activity. As their husbands circled the Earth, Patricia McDivitt (left on right-hand picture) and Patricia White spoke with them from mission control. When White (left on left-hand picture) and McDivitt landed, they spoke from the aircraft carrier* Wasp *to President Lyndon Johnson.*

[Center and bottom] *Gemini 4 brought fame to White and McDivitt. During the same month as their historic flight, they met Yuri Gagarin (front row, left) and Vice President Hubert H. Humphrey (second from left) at the Paris International Air Show and received congratulations from President Johnson and NASA Administrator James Webb (far right).*

Gilruth's team became too large to be accommodated in their ad hoc quarters at Langley. In order to design and integrate the gargantuan components destined for the Apollo program, the Space Task Group needed proximity to a large urban labor pool and supporting industries, as well as access to major shipping lanes. Webb found the answer when he inaugurated the Manned Spacecraft Center in Houston, Texas, a region connected to the Gulf of Mexico via the Galveston ship channel. This new entity not only assumed responsibility for all human missions into space and served as the astronaut training center, but also developed the spacecraft necessary for the missions. It opened in 1962 and became known in 1973 as the Lyndon B. Johnson Space Center. Meanwhile, Webb also needed to satisfy the Moon program's requirement for gigantic launch facilities. Clearly, NASA had outgrown its tenant relationship with the USAF on Cape Canaveral, Florida, one in which the Air Force provided launch and tracking for the early NASA missions. Webb, therefore, won congressional approval for NASA to purchase 111,000 acres (44,900 ha) on Merritt Island, adjacent to the USAF facilities. This immense parcel offered easy waterway access for the transport of massive rocket sections, yet sufficient isolation from inhabited areas to absorb the roar of launches and even the blast of potentially violent explosions. Famed Launch Complex 39, the take-off point of the nation's lunar expeditions, rose on Merritt Island, as did the thirty-six story Vehicle Assembly Building where technicians lashed together the components of America's Moon vehicles. This spaceport, called the Launch Operations Center, eventually bore the name of President John F. Kennedy. Finally, to enable safe ground tests of NASA's biggest rockets, laborers hewed an immense tract of bayou, which opened in 1961 as the Mississippi Test Facility (renamed the John C. Stennis Space Center in 1988). Surveying, clearing, and constructing these three sites cost the taxpayer about $2,200,000,000.

Inspired by an intense superpower rivalry that compelled both sides to pursue their ambitions at a frenzied pace, many events associated with the early U.S. space program became telescoped, occurring almost concurrently rather than sequentially. For example, as planning for Project Gemini and the subsequent Moon program unfolded in Washington, D.C., Huntsville, Alabama, Cape Canaveral, Florida, and Houston, Texas, NASA engineers in California laid the ground for an entirely new method of spaceflight.

It stemmed from an insight by Ames Research Center physicist H. Julian Allen who, in 1950, published the blunt body hypothesis. Allen theorized that as it re-entered the Earth's atmosphere from space, an object with a rounded, compact shape—subject to pressure drag—would protect

← As with Mercury, the Gemini missions flew on the back of Air Force missiles. The Titan 2 shown here lifted off with Lovell and Borman on December 4, 1965 (61628).

↓ Gemini 7 resulted in another space milestone. Astronauts James Lovell (left) and Frank Borman embarked on a complicated mission to rendezvous with another spacecraft, a feat essential for the later success of Apollo.

itself from burning up by heating the air surrounding it rather than its own skin. Pointed vehicles, in contrast, became intensely hot, experiencing frictional drag. The contours of the Mercury capsule, protected by a heat shield, conformed generally to Allen's basic concept. Later in the decade another Ames researcher, Alfred Eggers, proposed a practical vehicle that conformed to the blunt body shape: a wingless cone, cut in half lengthwise, with rear vertical fins, elevons for directional control, and a cockpit canopy. In wind tunnels it showed early promise for stable flight at hypersonic speeds, but also instability in the subsonic range. While Eggers and his team tested the design of this "lifting body" (named for its unusually high lift-to-drag ratio) at Ames, engineer R. Dale Reed at NASA's Flight Research Center on Edwards Air Force Base became intrigued with the design as a possible full-sized flight research project. The X-15 had been the first aircraft to enter space; Reed wondered whether the lifting body might represent the first spaceplane capable of flying back to a runway.

Initially—during late 1961 and early 1962—Reed (assigned to the X-15 program) auditioned several lifting body designs during his off hours as a model airplane enthusiast. The results encouraged him to see whether his bosses at the Flight Research Center might approve full-scale lifting body tests. It took persistence to translate the theories of Allen, the experiments of Eggers, and his own scale model data into an actual flying project. Reed needed to persuade three important figures to back his concept: center director Paul Bikle, Alfred Eggers himself, and prominent research pilot Milton O. Thompson.

Bikle balked at first, but relented after the endorsements of Eggers and Thompson. Bikle also hesitated to let anyone actually fly a lifting body, until Thompson won him over to piloted missions. Meanwhile, Reed designed a prototype plane called the M2-F1 (for Modification 2, Flight Version 1). Local artisans fabricated it from plywood and metal tubing, and outfitted it with off-the-shelf parts and a primitive flight control system. In contrast to the vast expenditures inherent in Mercury,

Gemini, and the Moon project, NASA paid just $5000 to build the first lifting body.

The resulting vehicle—an awkward, gumdrop-shaped machine—flew not on its own power but by being towed, first behind a souped-up 1963 Pontiac Catalina automobile, later by a C-47 aircraft. Milt Thompson flew it for the first time on March 1, 1963. Unfortunately it proved to be virtually uncontrollable, even though it barely lifted off the ground. Reed and his cohorts tested it repeatedly in the Ames 40 by 80 foot (12 by 24 m) wind tunnel before modifying the M2-F1's flight controls and once more trying it aloft in spring and summer 1963. Thompson reported a remarkable improvement, and it performed well on its first free flight (released from the C-47 towline) in August. But lateral roll control proved to be a recurring problem aboard the next generation of lifting bodies, the heavyweight, rocket-powered M2-F2, M2-F3, and the Horizontal Lander (HL)-10. In time, computerized stability augmentation systems and a middle fin between the rear vertical fins redressed the problem. According to the pilots who flew it, the last of the lifting bodies—the needle-nosed Air Force–NASA X-24B—flew as well as an F-104 fighter. Thus, even as the United States was gearing up for the Moon shots, the concept of a plane capable of maneuvering in space and returning home to a piloted landing underwent a decade of testing under Western skies, eventually proving itself fit for adaptation in the years ahead.

Meantime, the main objectives of the early American space program went forward. Gemini proceeded, but not without setbacks. Engineers discovered that the *Titan II* launch vehicle suffered from longitudinal oscillations that took time to remedy. They also found themselves redesigning leaking fuel cells. Transforming Agena from an Air Force system to a NASA rendezvous and docking vehicle took longer than anticipated. A hoped-for paraglider touchdown on land had to be abandoned in favor of a splashdown at sea like Mercury. Costs trebled from the original estimates.

The Gemini flight program began with two unmanned

THE X-15 HAD BEEN THE FIRST AIRCRAFT TO ENTER SPACE; DALE REED WONDERED WHETHER THE LIFTING BODY MIGHT BE THE FIRST SPACEPLANE CAPABLE OF FLYING BACK TO A RUNWAY.

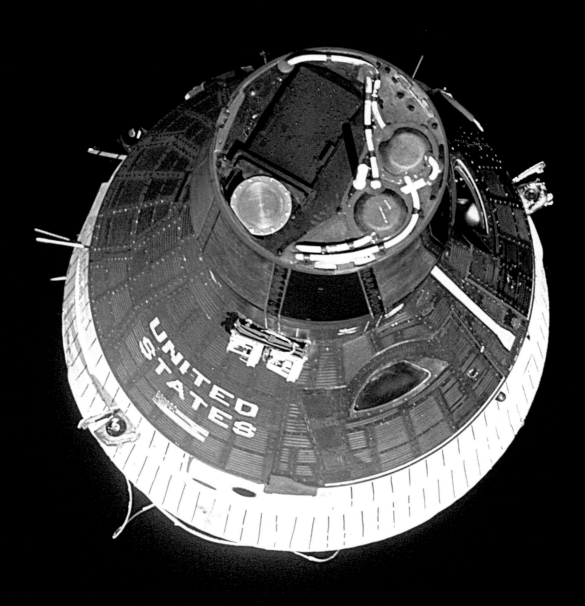

[Opposite and top row] *NASA mission planners decided to use two Gemini capsules for the first rendezvous. Accordingly,* Gemini 6, *crewed by Walter Schirra and Thomas Stafford, provided the rendezvous target. The two capsules approached on December 15 and maneuvered as close as 9 feet (2.7 m) from each other. The photographs opposite and top right show* Gemini 7, *the one above shows* Gemini 6.

[Right] *The astronauts of* Gemini 8—*Neil Armstrong and David Scott—tried the more problematical maneuver of docking with another spacecraft, chosen to be the Agena second stage rocket. Armstrong and Scott succeeded in docking with the Agena in March 1966, but their own capsule began tumbling, which Armstrong corrected by firing the re-entry rockets. Only with* Gemini 10 *and* 11 *did the techniques of docking in space become fully realized. Pictured, the large Agena vehicle as seen from* Gemini 8.

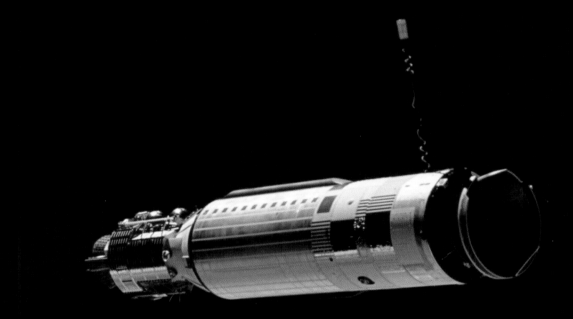

111

↑ *Armstrong and Scott, recovered in the Atlantic by the USS* Leonard F. Mason. *The Gemini capsule, with hatches open, rests on the spacecraft's yellow flotation collar, deployed to combat high seas.*

orbital missions. Then, in March 1965, the first human cargo went aloft when Gus Grissom and John W. Young prepared for later rendezvous attempts by firing thruster rockets that changed the profile of the orbits. During the next launch (June 1965) astronaut Edward H. White left the capsule for the first American space "walk," or extravehicular activity (EVA). *Gemini 10* and *11* achieved the crowning result of the post-Mercury flights by successfully rendezvousing and docking with Agena. Finally, by the last Gemini mission, in November 1966, astronauts endured not just days but up to two weeks in space, further unlocking the secrets of human physiology in weightlessness. In large and small ways, Gemini advanced the Moon project technically, institutionally, and experimentally.

Still, no one could say what chance the nation stood of realizing John F. Kennedy's goal. The president did not live to see the attempt: he was assassinated on November 22, 1963 during a motorcade in Dallas, Texas. If anything his death added momentum to the project. Another great loss occurred when Hugh L. Dryden, so instrumental to the founding of the U.S. space program, died in December

1965, well before Americans touched the lunar surface.

Like Mercury and Gemini before it, the planners of the Moon project selected a name from the pages of mythology. They chose Apollo, who (as the son of the great god Zeus) stood at the apex of the religious ziggurat, representing the Sun, music, poetry, prophecy, and agriculture to the ancients. But far from being celestial, the concerns of those leading the lunar expedition became entwined in a series of baffling technical and bureaucratic dilemmas: How to arrive at the Moon, land on it, and return safely? How to analyse the lunar environment before visiting it, so that the astronauts did not risk loss of life? How to manage the technical complexities of Apollo? How to conceive and fabricate an entirely new family of rockets capable of developing the immense thrust necessary to power Apollo? How to design a three-person spacecraft that enabled the astronauts to accomplish their missions but afforded them at least minimum comforts during their 480,000 mile (772,000 km) round-trip journey to and from the Moon? How to train human beings to withstand the physical as well as the psychological rigors of Apollo? Finally, how to devise lunar experiments that accomplished something useful yet did not impose undue weight penalties on the carefully calibrated limitations on lift capacity?

The first and perhaps most vexing problem was that of achieving consensus among the many strong personalities in NASA on how to approach the extraordinary technical conundrum of arriving at the Moon, landing on it, and leaving it. At NASA Langley, where Robert Gilruth presided over human spaceflight, three main contenders vied for the prize during 1961 and early 1962. The first method, Direct Ascent, was favored by Wernher von Braun and his team. Its adherents liked its simple approach: launch a large space vehicle on the back of a new and gargantuan Nova rocket (a von Braun concept, capable of 40 million pounds/18 million kg of thrust), land it directly on the lunar surface, and return home in a flyable portion of the spacecraft. Its detractors felt it posed too many technical obstacles and cost far too much. A second idea, known as Earth-Orbit Rendezvous, hinged on employing already planned rocket technology to orbit modular pieces around the Earth, assemble them in space, fuel the constructed spacecraft aloft, and fly it to the lunar surface. Less expensive and less daring technically than Direct Ascent, it attracted the support of von Braun after he abandoned the first option. However, at least during the early debates, the backers of the Direct Ascent and the Earth-Orbit Rendezvous options failed to agree upon a method of actually putting the astronauts on the lunar surface.

The final proposal for the voyage to the Moon, called Lunar-Orbit Rendezvous (LOR), encountered the most

[Right] *Although the early American efforts to lift human beings into space occurred aboard military launch vehicles, the success of Apollo depended on rocketry designed specifically for spaceflight purposes. Wernher von Braun personified these developments and displayed a deft touch at winning public support. In this picture von Braun (right) receives the Smithsonian Langley Medal, presented in June 1967 by Dr. Fred Whipple, director of the Smithsonian Astrophysical Observatory.*

[Bottom] *After months of heated controversy within NASA about how to land astronauts on the Moon and return them home safely, the path proposed by a relatively obscure scientist at Langley, named John Houbolt, was to prevail over the viewpoints of von Braun, Maxime Faget, and others.*

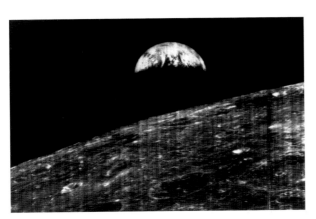

←⋯ ↑ *Later in the 1960s, the Lunar Orbiter program sent spacecraft to orbit the Moon and photograph it. On August 23, 1966, Lunar Orbiter 1 captured the world's first image of the Moon taken from a nearby vantage point, showing the Earth as a crescent over a broad sweep of lunar landscape.*

resistance, at least in the early debates at Langley and NASA headquarters. It originated in several minds at once, but found its best and most ardent expression in a relatively obscure structures scientist at Langley named John C. Houbolt. Unlike the two other alternatives, Houbolt's plan offered the simplest, yet the most risky choice. It also offered some tantalizing advantages over the others. In Houbolt's scheme, a spacecraft would take off from Earth aboard the standard Apollo launch vehicle and fly directly to the Moon. Once in orbit around the Moon, a small lander with a crew of two would detach itself from the main spacecraft and touch down on the lunar surface. Upon completion of its mission, this lunar module (as it came to be called) would lift off from the Moon and dock with the orbiting spacecraft for the return to Earth. On the face of it, Houbolt's proposal seemed to

have one great flaw: it appeared to imperil the lives of the astronauts. Unlike the Earth-Orbit Rendezvous plan, mission controllers (some 240,000 miles/386,000 km away during the complicated rendezvous and docking maneuvers over the lunar surface) had few options to save the lives of the astronauts in the event of a significant incident. But it also offered benefits, most important of all a tremendous reduction in Earth payload weight. Houbolt calculated the savings to be between factors of two and two and a half.

Houbolt faced a steep climb to convince the agency to adopt his concept. Far more noted individuals such as Abe Silverstein at headquarters, von Braun at Marshall, and Maxine Faget at Langley expressed unambiguous distaste—indeed, almost contempt—for the Lunar-Orbit idea. Not surprisingly, in the face of such powerful

opposition, during the second half of 1960 and most of 1961 much of the agency aligned itself against him. But Houbolt, who had conceived of the idea in an almost revelatory moment during the summer of 1960, pursued it with a zeal and a tenacity that at times seemed obsessive and unreasonable. Time and again, those who heard him present his concept left feeling that they had heard something plausible, but a technological long shot.

Wernher von Braun's eventual conversion to Houbolt's position— based on the Langley scientist's elaborate and clear-headed calculations that proved LOR's efficiency beyond any reasonable doubt—won the day. At a crucial meeting in June 1962, von Braun announced that Marshall had decided to throw its weight behind LOR. Houston and Langley then joined hands in the decision, and in July 1962 Administrator Webb—who with President Kennedy's

↑ *While the Rangers and Lunar Orbiters were undergoing development at JPL, on the other side of the country the rocketeers at Marshall Space Flight Center were static testing the first stage of the* Saturn I *launch vehicle, consisting of a bundle of eight H-1 engines.*

science adviser Jerome Weisner actually favored Direct Ascent—deferred to his staff and announced to the public the method by which the astronauts planned to make their approach to and return from the Moon. This decision by Webb settled one of the most contentious aspects of the Apollo program.

But this solution still left many other questions unresolved. For instance, no one seemed to know for certain just what awaited the astronauts when they walked out of John Houbolt's lunar module. How much dust would be disturbed? Would it inhibit visibility during touchdown? Would it be so fine that it might envelope the spacecraft? Did the lunar surface have enough firmness to support the weight of the spacecraft? Did radiation pose a threat to the astronauts' health? Did the conditions on the Moon render communications difficult or impossible? In short, scientists did not yet know enough about the environment and geology of the Moon to be able to predict its interaction with human activity.

Three projects, all managed by the Jet Propulsion Laboratory (JPL) in Pasadena, California, attempted to illuminate these questions. The first, Project Ranger,

⟨---- ----⟩ During 1961, the Saturn's H-1 engines underwent exhaustive testing, including immersion in salt water off Cape Canaveral and ground firings at Rocketdyne's main plant at Canoga Park, California. The eight H-1 powerplants mounted on Saturn I collectively produced 1.5 million pounds (680,000 kg) of thrust.

actually began during the late 1950s in the face of Soviet lunar explorations. In all, nine Ranger spacecraft flew— and most failed. But *Ranger 7*, which purposely hit the Moon on July 31, 1964 (nearly three years after *Ranger 1*), proved to be a colossal success. It approached the Moon precisely on target and fifteen minutes prior to impact its television cameras began transmitting the first of 4316 photos, many thousand times the resolution of those taken from Earth. The resulting analysis convinced JPL scientists that Apollo could indeed land safely on the lunar terrain. They confirmed their suspicions early the following year when *Ranger 8* sent home 7137 images before crashing into the Sea of Tranquility, where Apollo planners hoped to land the astronauts. The series ended with *Ranger 9*'s cameras taking over 5000 pictures as it descended into the crater Alphonsus, of geological interest but not related directly to Apollo.

Meanwhile, in 1960, the Lunar Orbiter program started, another JPL initiative designed to scan the Moon's surface for the optimal Apollo landing sites and, like Ranger, not originally conceived as a handmaiden for Apollo. It encountered a better success rate than Ranger, flying five satellites successfully around the Moon from August 1966 to August 1967. The first, *Lunar Orbiter 1*, mapped nine potential touchdown points for the astronauts, seven subsidiary sites, and one spot for the landing of the final pre-Apollo probes, called Surveyor.

Unlike the hard fate of Ranger, Surveyor promised the scientists at JPL a far more ambitious soft landing, its pictures continuing to stream back to NASA's tracking

←···· *The mighty* Saturn V *rocket's first stage fuel assembly being mated to the liquid oxygen tank in December 1964 at Marshall Space Flight Center. The five F-1 engines on the first stage each developed 1.5 million pounds (680,000 kg) of thrust.*

↓ *During a visit to Cape Canaveral, President Kennedy (right) points upward, Wernher von Braun (center) follows his glance, and NASA Associate Administrator Robert Seamans (left) looks on. The president did not live to see the firing of the* Saturn V, *but the developmental* Saturn I, *capable of 1.5 million pounds (680,000 kg) of thrust in its first stage, did fly a number of times between October 1961 and his death.*

stations after it settled down on the Moon. Conceived originally as a means to gather scientific data, under the demands of Apollo it became an engineering project that centered on a sophisticated imaging system. Amazingly, the designers of Surveyor achieved their ambitious objectives on the first try. On June 2, 1966, the 9 ft 10 ins (3 m) tall, three-legged spacecraft descended slowly toward the Moon, guided by Doppler radar linked to the on-board computer. Resting on the Ocean of Storms, it transmitted over 11,000 images before being shut down in January 1967. It also gleaned pivotal data about the density of the lunar surface, the degree of radar reflection, and the temperature ranges likely to be encountered by the astronauts. Thus, in the aggregate, Ranger, Lunar Orbiter, and Surveyor alleviated some of the darkest fears of the Apollo mission planners.

But other fears remained. One of the greatest involved the management of an enterprise the size of Apollo. Except perhaps for the digging of the Panama Canal and the Manhattan Project, Americans had never committed themselves to a government sponsored program of such magnitude. In the case of Apollo, James Webb found that not long after President Kennedy issued his challenge, Apollo began to fall behind schedule, raising the prospect of delaying the first lunar landing until 1971 and missing the late president's timetable.

As a consequence, in spite of strong resistance from the von Braun group in Huntsville, as well as from Robert Gilruth, Webb gradually embraced the practice of "all-up" development practiced by the U.S. Air Force, and specifically by the father of USAF ballistic missiles, General Bernard A. Schriever. In contrast to the slower, sequential development method perfected by von Braun and others—a conservative and effective procedure by which each component of a total system underwent testing and verification in turn—the faster all-up or concurrent approach enabled program managers to integrate components and test them as a unit, even to test them for the first time in flight.

Webb found the all-up method irresistible, as the lunar program was now as desperate for time as it was for money. A former director of the Bureau of the Budget, he valued efficiency and realized the dire political consequences if Apollo were to miscarry due to cost

overruns, to technical failure, or to tardy completion. To avert such occurrences, he persuaded the Air Force to detail to NASA a distinguished program manager skilled in the ways of concurrent development. General Samuel C. Phillips, a man of broad management experience, became deputy director in January 1964, and the following December the director of Project Apollo. Phillips transfused NASA program management with USAF experience by importing more than four dozen Air Force officers trained in the processes of the ballistic missile program. They imbued a wary NASA engineering corps—most of whom cut their teeth on the highly individualistic NACA research style preferred by George Lewis, Henry Reid, and also Hugh Dryden—with such principles as computerized data storage and retrieval, tight control over redesign, close supervision of cost and the flow of resources, careful monitoring of schedules, and a hierarchy of formal management reviews. By halting, or at least slowing down, the natural inclination to revisit earlier technical decisions, Phillips enabled the Apollo engineers to establish a baseline on which to estimate outlays and deadlines. If events proved them wrong, the management reviews allowed them to make appropriate adjustments. Perhaps most important of all, concurrent engineering began to be practiced by government and contractors alike.

← A technician stands atop the white room (the area through which the astronauts entered their capsule) as this Saturn V is prepared for the first Apollo mission with human beings aboard.

↓ Dr. Wernher von Braun (tenth from left) and his German colleagues confer with General John B. Medaris (in uniform) at the U.S. Army Ballistic Missile Agency (ABMA) just before his retirement in 1959. Medaris—a tough, plainspoken leader—assumed command of ABMA in 1956 and by his departure had roughly trebled its staff to about 5000.

↑ *Robert Gilruth (far right) introduces the crew of* Apollo 1 *to the press in March 1966: (left to right) Roger Chaffee, Edward White, and Gus Grissom. All three would die ten months later when a fire in their capsule broke out on the launch pad.*

Despite these successes, Phillips failed to sweep away all vestiges of the less structured development practices embraced traditionally by the NACA. Researchers at Marshall argued that concurrency transferred decision-making from the informed expert on the scene to managers at far away headquarters. Many other NASA engineers, nurtured in the NACA research tradition, sided with the von Braun group. Certainly, Webb and Phillips succeeded in getting a better grip on Project Apollo's resources and engineering talent, a victory with long-term consequences for how NASA was to manage itself after the end of the Moon program. Nevertheless, the historic NACA and NASA pattern that inspired individuals to pursue independent lines of research with considerable autonomy continued to coexist beside the space agency's newer management practices.

The juxtaposition of these two, distinctly different approaches manifested themselves plainly in the conception, design, fabrication, and testing of the mighty Saturn rocket, the Apollo launch vehicle. Pressured to yield to Sam Phillips's elaborate reporting practices and the "all-up" testing philosophy, the rocketeers at Marshall adhered to the more familiar engineering style wherever possible, but conformed to the demands made from above as necessary.

Von Braun's system of "Monday Notes" exemplified the step-by-step research process based on engineering knowledge, as opposed to management control. Each week, subordinates two levels below him sent one page summaries of their weekly activities and problems. Von Braun made extensive marginal notes on them, bound them together, and circulated them back for all to read. The Monday Notes served Marshall well. It inhibited von Braun's immediate subordinates from concealing bad news or manipulating events while keeping him in the dark. Moreover, the distribution of the notes with his marginalia showed the staff exactly what he wanted and at the same time informed everyone what their colleagues had been doing. As a whole, the Monday Notes helped integrate the efforts of the Marshall staff during the intense development of the Saturn rockets, and encouraged problem solving across organizational lines.

Saturn actually began even before NASA won control over the Army Ballistic Agency and the von Braun team. By then, the German émigrés had already laid the foundations for the first generation, known as *Saturn I*, the first stage of which would consist of eight H-1 engines, derived from the existing propulsion systems of the Thor and Jupiter missiles. To succeed on the *Saturn I*, the Thor and Jupiter engines required extensive modifications. In the end, the von Braun team managed to raise the H-1's output to 188,000 pounds (85,270 kg) of thrust each on the *Saturn I*. The second stage consisted of six RL-10 engines from the Centaur launch vehicle (researched at the NACA's Lewis Laboratory and fabricated by Pratt and Whitney), which collectively added another 90,000 pounds (40,800 kg) of thrust to *Saturn I*. The *Saturn I* flew ten times between late 1961 and mid-1965. It lifted the first stage alone in the initial four firings, and for the remaining six united with the RL-10s, sending three Pegasus meteor detection satellites and three Apollo test capsules into orbit. Although the RL-10s needed the utmost care in handling, being powered by the highly explosive propellants liquid oxygen and liquid hydrogen, all ten flights of *Saturn I* proved to be successful.

As *Saturn I* underwent testing, the Marshall researchers concentrated on the next rung in the Apollo launch ladder, the *Saturn IB*. Like the *Saturn I*, the *IB* had eight H-1s, each now generating 205,000 pounds (93,000 kg) of thrust. The new J-2 engine (second stage) produced 200,000 pounds (90,720 kg). In all, the *IB* had 1.8 million pounds, the *Saturn I*, 1.6 million. Fired with the second stage, the *Saturn IB* could lift out of the atmosphere a payload of up to 62,000 pounds (28,100 kg) enabling the complete Apollo capsule–lunar landing module combination to be orbited for testing in space. Only one human crew flew atop the *Saturn IB*: in October 1968 three astronauts flew 163 times around the Earth to test the Apollo equipment.

Finally, the team at Marshall devoted themselves to the

Samuel C. Phillips

Born in Springerville, Arizona, in February 1921, Samuel Phillips received a B.S. degree in electrical engineering from the University of Wyoming. He won distinction in World War II after earning his wings as an aviation cadet in 1943. He flew with the 364th Fighter Group of the 8th Air Force in England and received promotion to major after earning eight Air Medals, the croix de guerre, and two Distinguished Flying Crosses in combat. After the war, Phillips attended the University of Michigan and received the M.S. in electrical engineering in 1950. Most of Phillips's career involved technology management. He served for six years at the Engineering Division, located in Dayton, Ohio, rising to the rank of colonel as he became chief of the Air Defense Missiles Division, responsible for the development of the Bomarc and Falcon, among others. He then received an assignment to Air Force Systems Command's Ballistic Systems Division in Los Angeles, where General Schriever made him director of the Minuteman missile program with the rank of brigadier general. Here, Phillips burnished his reputation: he not only succeeded in making Minuteman operational, but accomplished it on schedule and below estimated costs. After serving with NASA during the 1960s, Phillips returned to the USAF as Commander of Air Force Systems Command during the early 1970s and retired in 1975. He died in 1990.

⟵ ····⟶ Many felt that President Kennedy's promise of a lunar mission by the end of the 1960s might go up in the smoke of Apollo 1. But Apollo 6, pictured here, took off in April 1968 to test the Saturn launch system. Apollo 7 sent the first Apollo crew into orbit in October 1968. Then in December 1968, Apollo 8 took the daring leap of attempting a circumnavigation of the Moon. Illustrated are the crews of Apollo 7 and 8 signing a commemorative document at the White House. Pictured, left to right seated: Apollo 7 astronauts Walter Cunningham, Donn Eisele, and Walter Schirra; and Apollo 8 astronauts William Anders, James Lovell, and Frank Borman. Standing: Charles Lindbergh, Lady Bird Johnson, President Johnson, James Webb, and Vice President Hubert Humphrey.

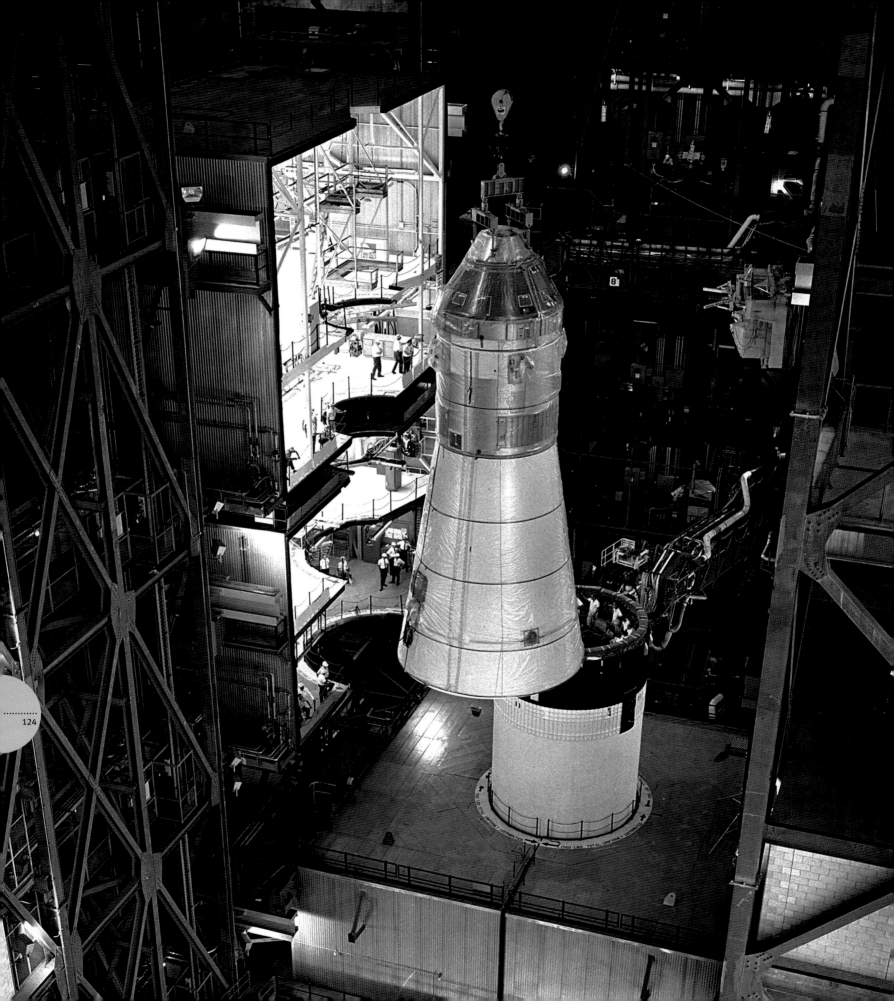

big technological leap: the immense *Saturn V* Moon rocket, without which the lunar project had no hope of success. If its 363 foot (111 m) stature and 9 million pounds (4.1 million kg) of total thrust failed to make an impression, its famed F-1 engine certainly earned the claim to gigantism. Begun as an Air Force project in 1955, NASA adopted the F-1 well before it had a launch vehicle to mount it on. In 1959 the space agency awarded Rocketdyne a contract to fabricate an F-1 capable of 1.5 million pounds (680,000 kg) of thrust, far more powerful than any American rocket at the time, but built from concepts that had already been proven. Once the powerplant became associated with Apollo, however, its role became more complicated. Launching the Apollo capsule to the Moon and back required 7.5 million pounds (3.4 million kg) of thrust in the first stage of the *Saturn V* alone, which entailed lashing five F-1s together to deliver adequate power. Thus, even though the F-1 used the standard propellant mixture of liquid oxygen and kerosene (the same mixture as the H-1 engines, although each F-1 produced 7.3 times the thrust of the H-1), its very size and power presented some formidable technical hurdles. For instance, project engineers needed to find new alloys and new construction techniques able to

withstand the intense heat generated by the F-1s.

The developmental challenges of the first stage of the *Saturn V* seemed mild compared to the anxieties of the second. Composed of five engines capable of 1.25 million pounds (567,000 kg) of thrust combined, its propellants consisted of the dangerous liquid oxygen and liquid hydrogen cocktail. In addition, the second stage of *Saturn V* fell behind schedule and required intense investigation (as well as additional funding), only narrowly avoiding the ignominy of being the components that thwarted President Kennedy's vision. The final stage of Saturn—designed to push the Apollo spacecraft and lunar module out of the atmosphere and towards the Moon—consisted of a single J-2 engine.

Before the first *Saturn V* flew, however, Apollo suffered a full-scale catastrophe. In the early evening of January 27, 1967, Commander Gus Grissom and astronauts Edward White and Roger Chaffee sat aboard the Apollo-Saturn 204 stack—better known as *Apollo 1*—on the Kennedy launch pad. Suddenly, as they practiced launch simulations, a flash fire erupted in the capsule. Fed by the pure oxygen that the crew breathed, the flames roared to life and the three men died of asphyxiation. The country and NASA reacted to the news with disbelief and horror.

⟨···⟩ The year 1968 proved to be crucial to Apollo. Not only did the space agency recover from the Apollo 1 *catastrophe by launching a string of successful spaceflights, but on the ground engineers succeeded in bringing together the main components required for the Moon missions. Seen here, a close-up of the docking system of the Apollo spacecraft, and a view inside the mammoth Vehicle Assembly Building at Kennedy Space Center, showing the mating of the command and service modules to the instrument unit on the* Saturn V.

---> *Astronaut and former research pilot Neil Armstrong at the Lunar Landing Research Facility at Langley Research Center, February 1969. Armstrong had previously flown the Lunar Landing Research Vehicle at the Flight Research Center, Edwards, California.*

↓ ↘ ↘ *A month before their historic trip around the Moon, the Apollo 8 crew (James Lovell, William Anders, and Frank Borman) pose in their space suits on a Kennedy Space Center simulator. Three days after leaving Kennedy, the American astronauts photographed the far side of the Moon. If this view lacked sufficient drama, Lovell, Anders, and Borman got the sight of their lives as they flew from behind the Moon: they saw the Earth five degrees above the horizon, 240,000 miles away.*

Many thought a tragedy might befall the space program one day; no one imagined that disaster would strike on the ground. A panel appointed by Administrator Webb investigated the calamity and found that a short in the capsule's electrical system had initiated the blaze. The subsequent report called for modifications of the spacecraft to enable the crew to escape, and also wanted the oxygen in the capsule reduced during ground tests. Although NASA complied quickly with the recommendations, Congress and the media tarnished Webb's reputation. The space program as a whole continued to enjoy broad public support, but NASA as an institution also suffered at the hands of the critics.

Meanwhile, the *Saturn V* Moon rocket had its hour of trial on November 9, 1967, the first Apollo launch since Grissom, White, and Chaffee had lost their lives. As if the moment lacked sufficient drama, on that date, the "all-up" testing philosophy imported from the Air Force Ballistic Missile program also got its acid test. The mission, known as Apollo 4, consisted of the launch of the complete Apollo–Saturn combination, never tested as a unit until this flight. Failure on the heels of Apollo 1 would render President Kennedy's end date of 1969 all but impossible. At 7 am on that day the rocket lifted off and flew perfectly, the best possible antidote to the gloom and pessimism engendered by Apollo 1. The following launch, Apollo 6 in April 1968, did less well. The second stage cut off

prematurely and the third stage failed to ignite on the second burn—but NASA declared it a success anyway. Indeed, despite the organizational innovations, the technical complexity, and the shortage of time, success crowned the Saturn. Among seventeen test flights and fifteen launches with humans on board, the Saturn rockets may have malfunctioned occasionally, but they never actually failed.

Hard as the Saturn development may have been, it represented only part of the difficulties encountered during the design and fabrication of Apollo. The project's spacecraft, as well as the lunar module (LM), also presented formidable obstacles. The spacecraft design and test, awarded to North American Aviation in November 1961, went on for almost seven years. The long development related in part to the complexity of the vehicle. Mission planners needed a machine that was capable of both orbiting the earth and flying to the Moon and back; of sustaining three human beings for more than two weeks; and of accommodating a service module (equipped with a main engine for the re-entry) that would store sundries such as life support equipment, fuel, maneuvering rockets, and oxygen. The long gestation also hinged on the calamity of the launch pad fire of *Apollo 1*. It took time to make subsequent modifications of the spacecraft in order to enable the escape of future crews and to reduce the dependence on oxygen in the

Virgil I. Grissom

Virgil Ivan "Gus" Grissom, born in April 1926, grew up in Indiana, the son of a Baltimore and Ohio railroad worker. He earned a B.S. in mechanical engineering from Purdue University in 1950, after which he joined the U.S. Air Force. Grissom served as an F-86 combat pilot during the Korean War and won the Distinguished Flying Cross. He remained in the USAF after the war, attending the Air Force Test Pilots School, after which he reported to Wright Patterson Air Force Base, Ohio, where he flew new jet engines. After being selected for astronaut training for Project Mercury, he (like the other seven Mercury astronauts) became a celebrity. He also became the second American in space when he flew *Liberty Bell 7* in July 1961, narrowly escaping death when the emergency hatch on his capsule opened unexpectedly after his splashdown in the Atlantic Ocean, forcing him to swim free. In March 1965 Gus Grissom commanded the first Gemini mission, a two-person flight that he shared with John Young in which he had the distinction of being the first astronaut to maneuver a spacecraft as it orbited the Earth. Lieutenant Colonel Grissom died at the age of forty in the *Apollo 1* fire and his remains lie in Arlington National Cemetery.

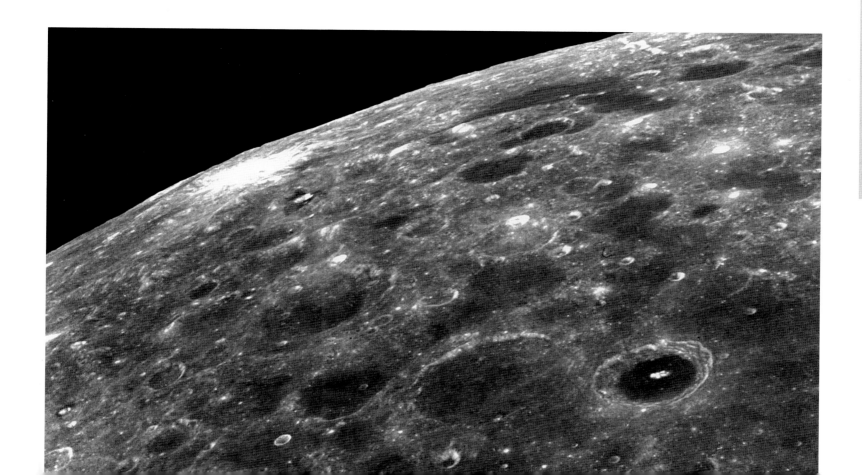

Neil A. Armstrong

Neil A. Armstrong (born in 1930) was only thirty-two years old when he was selected to the astronaut corps, but already had a wealth of flying experience. Born and raised in Wapakoneta, Ohio, Armstrong joined the Navy and served as an aviator during the Korean War, flying seventy-eight missions in combat aboard the F9F-2 fighter and received the Air Medal and two Gold Stars. He then earned a bachelor's degree in aeronautical engineering from Purdue University (and later a master's in aerospace engineering from the University of Southern California). In 1955 Armstrong decided to join the staff of the National Advisory Committee for Aeronautics at the Lewis Research Laboratory in Cleveland, Ohio. Later that year he transferred to the NACA's High-Speed Flight Station at Edwards, California, where, as a research pilot, he amassed 2450 flying hours aboard such varied aircraft as the F-100 A and C, the F-101, the F-104, F-105, F-106, B-47, and KC-135. During his career Armstrong also flew a number of experimental vehicles, including the X-1B, X-5, and the famed X-15 hypersonic rocket plane, which he piloted seven times up to a speed of Mach 5.74. Armstrong spent a total of eight days and fourteen hours in space. He retired from NASA as Deputy Associate Administrator for Aeronautics at headquarters in 1971, after which he became a professor of engineering at the University of Cincinnati. He also chaired a number of corporate boards. Armstrong served as the vice chair of the Challenger Presidential Commission in 1986. Among many other honors, he won the prestigious Collier Trophy in 1969.

spacecraft's atmosphere. In the end, the last checkout of the vehicle prior to lunar flight occurred between October 11 and 22, 1968, during the flight of *Apollo 7* and its three crew members.

The lunar module—a component of the Apollo spacecraft—represented perhaps the most problematical portion of the Apollo hardware. Started late, perpetually behind schedule, and far more costly than anticipated, the LM needed to achieve the most daring technological feat if the astronauts had a chance of returning from the Moon. It embodied the genius—and the weakness—of John Houbolt's plan for the lunar voyage. The lunar module needed to achieve two goals. First, to detach itself from the *Apollo* spacecraft as it circled the Moon and descend to a precise spot on the lunar surface. Then, after the astronauts completed their mission, the LM needed to ascend from the Moon, rendezvous with the orbiting ship, and reconnect for the trip back to Earth. To succeed, the engines on both halves of the LM had to work flawlessly, and engineers involved in the project fought constantly to resolve control, guidance, and maneuverability problems.

The Grumman Corporation won the contract to build the LM and stayed on the project from late 1962 until the beginning of 1968 (when a *Saturn V* test orbited the lunar module) and beyond. Grumman had some big problems. During June 1967 the Kennedy Space Center took delivery of the company's first lunar module. When it failed the leakage tests miserably, Kennedy's imposing and hot-tempered Director of Launch Operations, Rocco Petrone (an Army lieutenant colonel and a former West Point football player), called in Grumman's local representative and dressed him down mercilessly. "What kind of two-bit garbage are you running up in Bethpage?" he demanded. "What kind of so-called tests did they do in New York

Neil Armstrong (far left) trains in the Apollo lunar module mission simulator at Kennedy Space Center. Armstrong was known as a no-nonsense, reserved personality when he worked at the NASA Flight Research Center, and maintained this demeanor when he joined the Apollo program. Michael Collins (center left) practices docking techniques inside a command module mockup, specifically in the tunnel leading to the lunar module. After his career as an astronaut, Collins became the first director of the Smithsonian Institution's National Air and Space Museum. Astronaut Buzz Aldrin (left) takes photographs during training aboard a KC-135 aircraft. Aldrin became a leader in the non-governmental space exploration movement once he retired from the astronaut corps.

An official portrait of the astronauts of Apollo 11: left to right, commander Neil Armstrong, command module pilot Michael Collins, and lunar module pilot Edwin Aldrin, taken at Johnson Space Center two months before their lunar mission.

before sending this wreck to us? . . . They had better get this fixed, and fast! Your name is mud around here until they do." The Grumman man conveyed Petrone's sentiments to his plant with equal force, and the matter resolved itself after a shake up at the company.[2]

One other element of Apollo needed to be readied before the launch: the human beings to be sent aloft. As early as January 1959, Gilruth and his team at Langley had announced seven minimum criteria for admission to the astronaut corps: a college degree in the physical sciences or engineering; graduation from a military test pilot school; at least 1500 hours' flying time, preferably in jet fighters; a maximum age of forty; a maximum height of 5 ft 11 ins (1.8 m); excellent physical condition; and a psyche suited to the perils of spaceflight. After putting the initial candidates through many levels of physical and psychological screening, the seven final Mercury astronauts became known to the public. The adulation they encountered astounded even the NASA authorities. John Glenn alone received more than 350,000 pieces of mail and 1400 requests for appearances just after he orbited

At last, on July 16, 1969, the Apollo 11 mission lifted off from Kennedy Center Launch Complex 39A. After the launch of Apollo 11, a palpable sense of relief went through mission control at the Johnson Space Center. Pictured here are Wernher von Braun (wearing binoculars); Charles Matthews, Deputy Associate Administrator for Manned Space Flight (to von Braun's right; George Mueller, Associate Administrator for Manned Space Flight (to von Braun's left); and Lieutenant General Samuel Phillips, Director of the Apollo program (in short-sleeved shirt).

the Earth. In time, the individuals chosen to fly into space regained some control over their lives by organizing the Astronauts Office at the Manned Space Center.

The first flight training program, developed by Navy psychologist Robert Voss in April 1959, consisted of rigorous ground school and flight test phases. Significantly, among the second group of astronauts (selected in September 1962) two of the nine did not hold military commissions, since the requirement for test pilot school had been waived. Both Elliott See and the subsequently more famous Neil A. Armstrong benefited from this revision of the rules.

Although Apollo survived unending technical, administrative, and fiscal adversities, its flights seemed imbued at times with a charmed life, even when disaster loomed. Late in 1968 its program managers took a big gamble in order to recover time lost in the aftermath of Apollo 1. Rather than attempt a low Earth orbit to test hardware, Manned Spacecraft Center senior manager George Low, abetted by Apollo program manager Samuel Phillips, argued successfully to push the program forward and make a circumlunar flight. Between December 21 and 27, 1968, Apollo 8 flew to the Moon, circled around its "dark

side," and on Christmas Eve astronauts James Lovell, Frank Borman, and William Anders transmitted an image of our planet emerging from darkness as no human had ever seen it. They read aloud from the Book of Genesis, "And God created the heavens and the Earth . . ." The risk of Apollo 8 paid off. On March 3 the following year Apollo 9 tested an extravehicular activity (EVA) contingency transfer between the LM and the Apollo spacecraft.

Then, on July 16, 1969—not long before President Kennedy's almost unattainable goal lost out to the clock— Apollo 11 launched from Kennedy Space Center. But after a perfect flight, serious problems materialized. As lunar module Eagle descended toward the Moon a little after 4 pm Eastern Time on July 20, pilot Armstrong had to cope with two unsettling events. First, the on-board computer indicated that it had become overloaded with commands. Second, and worse, despite the brilliant lunar surveying achieved by JPL, as the chosen landing site in the Sea of Tranquility came into view, Armstrong found to his dismay a large crater in an area expected to be flat. He decided to take command of Eagle from the computer. Armstrong's tense search for a suitable touchdown point almost came to nothing until he saw an unobstructed area and planted

[Above] *One of the few pictures of Neil Armstrong on the Moon, this one shows him looking into the modular equipment storage area on the lunar module. By this point, the American flag had already been planted on the Moon's surface.*

[Right] *A landscape similar to the one observed by Michael Collins as he awaited the return of Armstrong and Aldrin from the Moon's surface. The crew of Apollo 10 actually took this photo during their May 1969 mission to verify the separation of the command module from the lunar module, and their subsequent docking.*

[Opposite] *The* Apollo 11 *mission conducted a number of scientific projects. Against the background of the* Eagle, *Buzz* Aldrin *posed next to one of these investigations, the Solar Wind Composition Experiment.*

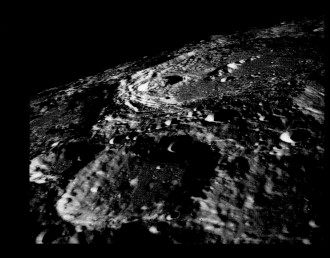

As the astronauts were acquainting themselves with the Moon, President Richard Nixon called to offer his congratulations for their achievement. Nixon declared it a time when "people on this Earth are truly one, one in their pride in what you have done . . ." After a period in quarantine and a press conference, Armstrong, Collins, and Aldrin embarked on a NASA-sponsored worldwide publicity tour. Immense crowds like this one showered the Apollo 11 astronauts with confetti and cheers wherever they traveled.

the spacecraft's feet on the lunar landscape. The lunar module had approximately thirty seconds of propellant remaining in the descent fuel tank, leaving very little time for maneuver.

Armstrong then took the celebrated "one small step" along with fellow astronaut Edwin "Buzz" Aldrin after climbing down the lunar module's ladder onto the dusty surface. For two and a half hours they walked the Moon as a television camera some 40 feet (12 m) from *Eagle* sent unthinkable images back to a stunned Earth. They took still pictures and pushed the U.S. flag into the surface. They collected 48 pounds (22 kg) of lunar material and experimented with ways to walk in 1/6 gravity. Then, the two men turned their backs on this new world, ascended the ladder, closed the hatch behind them, and departed for *Columbia*, which was circling above. The remainder of the mission unfolded predictably.

Thus, with about five months to spare, Americans succeeded in fulfilling the seemingly impossible objective of President John F. Kennedy. But, as Hugh Dryden, also so instrumental in this astonishing feat, once observed, the real value of space exploration occurred not in the heavens, but on Earth itself.

1. *Public Papers of the Presidents of the United States: John F. Kennedy, 1961*, quoted in Roger D. Launius, *NASA: A History of the U.S. Civil Space Program*, Malabar, Fla. (Krieger Publishing) 1994, p. 183.

2. Thomas J. Kelly, *Moon Lander: How We Developed the Apollo Lunar Module*, Washington, D.C. and London (Smithsonian Institution Press) 2001, pp. 165–66.

↑ The USS Hornet *picked up the Apollo 11 crew on July 24, 1969, just over 800 nautical miles (1480 km) southwest of the Hawaiian Islands. A Navy diver helps the astronauts. All of them wear biological isolation garments.*

With the news that Apollo 11 had been pulled from the ocean, officials at Johnson Space Center were overjoyed. Among the celebrants hailing the mission are (from left to right) Maxime Faget, George Trimble, Christopher Kraft, Julian Scheer (behind the others), George Low, Robert Gilruth, and Charles Matthews (between the flags).

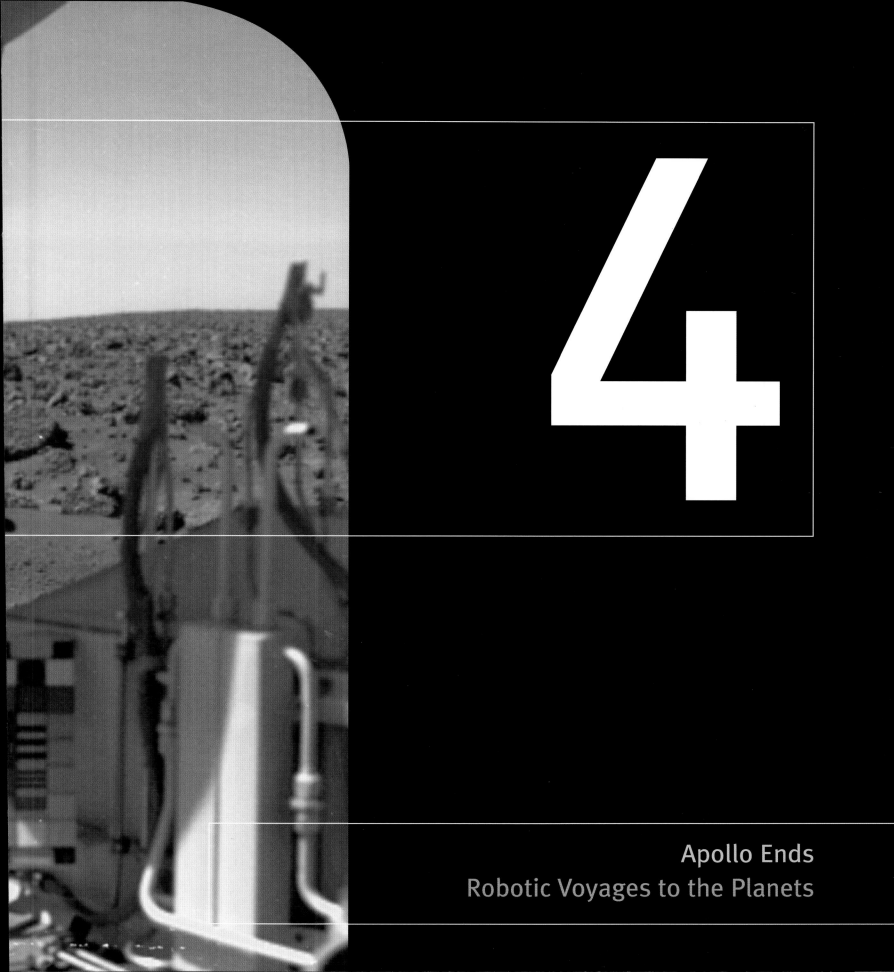

4

Apollo Ends
Robotic Voyages to the Planets

[Left and bottom left] *After a dusty and hazardous landing, Apollo 12 accomplished a great deal. Here, on November 19, 1969, Alan Bean descends the lunar module* Intrepid's *ladder to join commander Pete Conrad on the lunar terrain. The remains of the Surveyor lunar probe lay just a few hundred yards from this scene. The following day, a crew member poses with the Apollo lunar hand tools and carrier.*

[Bottom right] *The aircraft carrier USS* Hornet *cuts a path toward the Apollo 12 capsule as a helicopter from the vessel hovers over the point of splashdown. Navy seamen then hoisted up the astronauts, as well as a significant collection of geological samples collected from the Moon.*

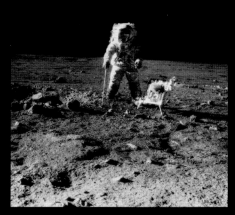

↑ ↗ *The Apollo 12 mission represented a second decisive success for NASA. The astronauts, (from left to right) commander Charles "Pete" Conrad, Richard Gordon, and Alan Bean, pose in their space suits about two months before the flight. Their Saturn V launch vehicle is shown being moved on its transporter from Kennedy's Vehicle Assembly Building to Launch Complex 39A—at a speed of one mile (1.6 km) per hour.*

↘ *President Richard Nixon (center), his wife Patricia (shielded under an umbrella by NASA Administrator Thomas Paine, far right), and their daughter Tricia (bottom row, second from left) watch pre-launch activities at the Kennedy viewing area prior to the Apollo 12 lift-off.*

Although the sojourn of Neil Armstrong, Edwin Aldrin, and Michael Collins represented the first (and the most tentative) steps on the Moon, it also signified the beginning of the end of Project Apollo. Despite the unprecedented adulation that now followed the three astronauts wherever they met the public, and as remarkable and unprecedented as the remaining six Apollo flights turned out to be, political and budgetary decisions in Washington, D.C., had already sealed the fate of the lunar program. As Apollo ran its course, however, an entirely different—but in its own way no less spectacular—series of space adventures lay ahead.

If Apollo 11 symbolized the fulfillment of President Kennedy's supreme Cold War gamble, Apollo 12 seemed routine, workmanlike, and almost devoid of political overtones. On November 19, 1969—four months after the first lunar mission—Commander Charles Conrad and Alan Bean guided the lunar module *Intrepid* onto the *Surveyor* Crater. But the touchdown had its adversities. Dust stirred by the descent engines obscured Conrad's vision and caused him to travel 600 feet (183 m) off target, which happened to be at the remnants of *Surveyor 3*. Conrad and Bean went on an EVA the first day, during which time they fought the fine powder on the Moon's surface (that stuck like concrete due to the lack of air molecules) as they set up an instrumentation suite known as the Advanced Lunar Science Experiment Package (ALSEP), fueled by electricity from heat generated by plutonium. After a rest, the two men took a long hike in an effort to determine human endurance in lunar gravity, assess the efficacy of their space suits, survey the landscape and the *Surveyor 3*, and collect geological samples. They walked over a mile (1.6 km) during a three and a half hour period and experienced little fatigue, although they did notice that their suits resisted bending at the thighs, causing

them to move straight-legged and to land flat-footed. Moreover, the internal pressure of the gloves fatigued their hands. After packing aboard the lunar material and pieces of *Surveyor 3* (to enable engineers to determine the degree of spacecraft deterioration), Conrad and Bean left the Moon and reunited with the orbiting *Yankee Clipper* piloted by Richard Gordon. The three then returned to Earth, splashing down in the Atlantic Ocean on November 24.

The flight of *Apollo 13* reminded the world of the razor-thin margin between safe and catastrophic spaceflight, conjuring a horrible flashback of the deadly *Apollo 1* fire. As he and his crewmembers John Swigert and Fred Haise launched on April 11, 1970, mission commander James Lovell felt a sense of confidence about the journey, having amassed 572 hours in space during Gemini 7, an American record at the time. Indeed, during the first two days of the flight, the vehicle performed better than its

↑ ↗ *Dressed in civilian clothes, astronauts Fred Haise, John Swigert, and commander James Lovell (shown left to right) are photographed in January 1970, about three months before their star-crossed voyage on* Apollo 13 *began. The launch vehicle and capsule stack left the Vehicle Assembly Building in December 1969 on its way to Kennedy's Pad 39A.*

predecessors. At almost the fifty-sixth hour, however, the crew heard a sharp bang and sensed vibration. Oxygen Tank 2 in the service module, having just been stirred mechanically, exploded, causing Tank 1 to fail. Lovell realized the gravity of the situation when he looked outside the capsule and noticed the contents from Tank 2 venting into space, resulting in a sphere of gas around the spacecraft. In one blow, Apollo's sources of air, water, electricity, and light had been compromised.

Engineers on the ground then made a life-saving improvisation. They told Lovell, Swigert, and Haise to abandon the command module and wait out the long trip to the Moon, around it, and back to Earth aboard the lunar module, which held barely enough electricity, air, and water to sustain the three men through the days ahead. By abandoning the capsule and holing up in the LM lifeboat, the fifteen minutes of power left in the command module could be preserved for the descent to Earth.

The entire world glued itself to radios and televisions, awaiting the outcome of *Apollo 13*, the fate of which seemed anything but certain. Mission controllers had to find a way to double the lunar module's normal forty-five hours of operating capacity. The capsule needed to be positioned so that the re-entry trajectory would enable the capsule to penetrate, rather than bounce off, the atmosphere. Commander Lovell needed to aim the capsule at the Earth not by use of a distant star but by the Sun (the debris hovering around the capsule made a standard reckoning impossible). The crew faced cold, hunger, thirst, and sleeplessness. Quite miraculously, *Apollo 13* plunged into the ocean safely on April 17, 1970. But paradoxically, rather than being perceived as a failure, the mission seemed to burnish NASA's image and prove that in a crisis, its scientists, engineers, and astronauts possessed the technical virtuosity to grasp life from the jaws of death.

The mission that followed Apollo 13 faced a hard task,

oxygen tanks (number two) exploded. As it drifted away, one of the astronauts captured this picture of the service module.

[Bottom row] *The tension inherent in the Apollo 13 drama appeared on the faces of those working to save the mission. During the crew's final twenty-four hours in space, many top level NASA officials gathered around the consoles at the Manned Spacecraft Center's mission control to offer guidance. Pictured bottom left (left to right), are Thomas McMullen, Shift 1 mission director; Dale Myers, Associate Administrator for Manned Spaceflight; Chester Lee, Apollo 13 mission director; and Rocco Petrone, Apollo program director. Alan Shepard's expression during the oxygen cell failure crisis illuminates the gravity of the moment.*

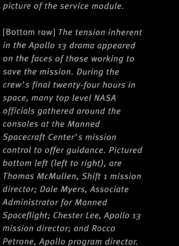

[Above and left] *An arm raised in triumph after the news of the splashdown. In the upper left, Manned Spacecraft Center Director Robert Gilruth lights up a cigar with deputy director Christopher Kraft. In the foreground, flight directors (left to right) Gerald Griffin, Eugene Krantz, and Glynn Lunney, show the elation of the moment. Also in the mission control center, Administrator Thomas Paine (bottom picture, center) applauds, joined by Lieutenant General Samuel Phillips (far left), Charles Berry (Director of Medical Operations) (second from left), and George Low (Associate NASA Administrator (far right).*

[Opposite] *At about noon on April 17, 1970, the* Apollo 13 *spacecraft floated through the atmosphere on parachutes on its way to splashdown. The crew of the USS* Iwo Jima *pulls the command module* Odyssey *from the waters of the South Pacific.*

Alan B. Shepard, Jr.

Born in East Derry, New Hampshire, on November 18, 1923, Alan Bartlett Shepard, Jr., chose a military career. After local schooling he entered the U.S. Naval Academy in Annapolis, Maryland. Upon graduation, Shepard served briefly in World War II, and afterward took flight training at Corpus Christi, Texas, and Pensacola, Florida, earning his Navy pilot's wings in spring 1947. A few years later he attended the Navy's Test Pilot School at Patuxent River, Maryland, and during the 1950s flew experiments related (among others) to aerial refueling and the new F8U Crusader. During 1959, NASA invited 110 American test pilots to volunteer for the space program and in April of that year Shepard became one of the original seven Mercury astronauts. Robert Gilruth designated him the prime pilot for Mercury, in effect making him the first American in space. Shepard piloted *Freedom 7* in a successful suborbital flight on May 5, 1961. Despite plans to continue in Project Gemini, Shepard reluctantly left the astronaut corps in early 1964 due to the discovery that he suffered from Ménière's syndrome, an illness that causes dizziness and nausea as a result of fluid buildup in the inner ear. Shepard decided to remain in the space agency, and became chief of the Astronaut Office. In 1969 he visited a physician in Los Angeles to try a new procedure to alleviate his illness. It succeeded, and in May 1969 the forty-six year old returned to flight status. Fellow astronaut from Mercury days Deke Slayton promptly named him to command *Apollo 14*. After the flight, he returned to the Astronaut Office, but left NASA (and retired from the Navy as a Rear Admiral) in July 1974. Shepard prospered in Houston in the construction and beer distribution businesses, and participated in charitable causes. He died in July 1998.

←⋯ *James Lovell testifies about the* Apollo 13 *misfortunes during an open session of the Senate Space Committee just a week after his homecoming. Administrator Paine is at the left.*

indeed. As well as demonstrating the effectiveness of new procedures that fixed the misbehaving oxygen tank, it required a proven commander to restore the nation's (and NASA's) confidence. Alan B. Shepard, of Mercury fame, proved equal to the challenge. He led Edgar Mitchell and Stuart Roosa to the Moon on January 31, 1971. The mission landed in the Fra Mauro region, originally selected for *Apollo 13*. Shepard exulted in his reversal of fortune, saying to ground controllers, "Al is on the surface, and it's been a long way, but we're here."[1] Shepard and Mitchell took two Moon walks, conducted seismic research, and deployed a pull-cart to carry their equipment and their samples (100 pounds/45 kg of rock in all). The Apollo 14 mission ended safely on February 9, 1971. The last of the lunar missions occurred over a period of a year and a half, beginning with Apollo 15 (July 26–August 7, 1971), then Apollo 16 (April 16–27, 1972), and ending with Apollo 17 (December 7–19, 1972). Unlike the earlier voyages, these resembled expeditions rather than quick forays, and featured a lunar rover, a vehicle that transported the astronauts over greater distances and more difficult terrain than was possible on foot.

Apollo succeeded exactly as President Kennedy hoped it might. It shattered the illusion of Soviet technological

←⋯ Apollo 14*'s crew had the unenviable task of attempting to restore the credibility of the lunar program after the ill-fated* Apollo 13. *The astronauts for the task included a face from the past, Alan Shepard, mission commander (center), accompanied by Edgar Mitchell (right) and Stuart Roosa.*

←⋯ ↗ *Alan Shepard suits up for the launch of* Apollo 14 *during the pre-launch countdown. He, Roosa, and Mitchell took off on January 31, 1971, the fourth time the United States attempted a lunar landing by the astronauts.*

The lunar module Antares, deflecting a beam of sunlight to the camera (below) and with a brilliant solar glare behind it (right). Tracks on the left side of the photograph resulted from the movement of the modularized equipment transporter away from the lunar module. Shortly after his first EVA, Alan Shepard erected the U.S. flag on the Fra Mauro Highlands. A shadow cast by the top of the lunar module is visible on the lower left of the photograph.

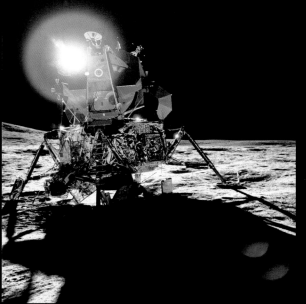

↑ Apollo 14 astronauts Stuart Roosa, Alan Shepard, and Edgar Mitchell (left to right) inside the mobile quarantine facility after the Navy picked them up aboard the USS New Orleans. Apollo 14 helped to restore confidence in the Apollo program after the near disaster of Apollo 13.

superiority; it showed that the American economic model had merit, simply because it produced such spectacular results; and it suggested that a democracy that operated openly might compete successfully against a totalitarian regime that conducted its affairs in secret.

Still, even though Apollo taught these lessons, it disappeared with surprising speed from the American scene. Even before the first Moon walk—in fact, as early as 1966—the budgetary tide had begun to run out on the U.S. space program. Once the race to the Moon had been won, the U.S.S.R. seemed less intimidating than in the years before the Moon launches. Moreover, pressed by such urgent political and fiscal priorities as the Vietnam War and President Johnson's "Great Society," members of Congress realized that the political constituency for spaceflight paled in comparison to the public interest in peace, war, and social justice. Finally, the impact of the assassination of President Kennedy—the early space program's great champion, who had banked his political survival on the Moon program—began to wane, freeing political leaders to formulate their own agenda without appearing to violate the spirit of the martyred president.

The rate of NASA's decline from full funding to virtual subsistence occurred gradually, but inexorably. Between 1959 and 1965, there had been a breathtaking increase: from an appropriation of $369,000,000 in 1959, to $1,829,000,000 three years later, and no less than $5,250,000,000 in 1965. Then, a ten year retrenchment began. Between 1966 and 1975, the NASA budget fell every year apart from one, declining from $5,175,000,000 in 1966 to $3,231,145,000 in 1975—an overall loss in annual funding of almost 40%, made worse by a sharp inflationary squeeze. At the same time, the number of federal employees on the NASA rolls fell by a third. Most of the reductions occurred during President Richard M. Nixon's administration. Although all the Apollo Moon landings happened during his presidency and he sympathized with the objectives of the space program, Nixon — under intense pressure to fund the Vietnam War and to increase social programs — reluctantly instructed James Webb's successor, Thomas O. Paine, to cut the agency's budgets to about $3,000,000,000 per year. Paine resisted mightily.

Rather than submit to Nixon's decision to make

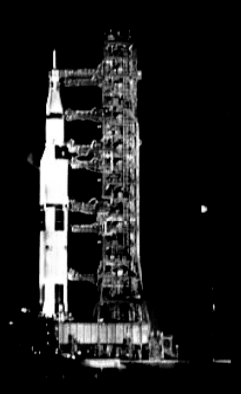

[Top row] Apollo 15 *crew (left to right) commander David Scott, command module pilot Alfred Worden, and lunar module pilot James Irwin. The night before the launch, the Apollo 15 stack is illuminated against the darkness .*

[Right and opposite] *The Lunar Roving Vehicle (on the right, with James Irwin outfitting it for the mission's first EVA) added an important new dimension to the last three Apollo missions, enabling long-distance treks and the collection of greater quantities of geological material. Turning the camera in another direction, the crew of the lunar module shot this striking photograph of the command–service modules in lunar orbit.*

[Opposite] *A driven perfectionist,
Eugene Krantz (Johnson's
Director of Flight Operations)
prepares his console at mission
control on the morning of the
launch of* Apollo 16. *On the Moon,
astronaut John Young snapped a
picture of Charles Duke at the
edge of Plum Crater, a depression
roughly 120 feet (36.5 m) in
diameter and 30 feet (9m) deep.
The* Rover *is in the background.*

[Above right] *The three astronauts
of* Apollo 16 *posed for an official
portrait before launch. Pictured
(left to right) are command
module pilot Thomas Mattingly,
commander John Young, and
lunar module pilot Charles Duke.
Taken from the ground, the
363 foot (110.5 m) tall Apollo 16
launch vehicle at Pad A,
Launch Complex 39, made an
imposing sight.*

[Right] *On April 27, 1972, the
commanding officer of the USS*
Ticonderoga *welcomed the
returning crew of* Apollo 16. *They
splashed down about 215 miles
(346 km) southeast of Christmas
Island in the Pacific Ocean.
Commander Young speaks from*

EVEN BEFORE THE FIRST MOON WALK—IN FACT, AS EARLY AS 1966—THE BUDGETARY TIDE HAD BEGUN TO RUN OUT ON THE U.S. SPACE PROGRAM.

↑ ↗ In the typical blast of fire and smoke, the last Apollo mission—number 17—leaves the ground on December 7, 1972. An on-orbit shot of two of the Apollo 17 crew, commander Eugene Cernan (left) and lunar module pilot Harrison Schmitt. Command module pilot Ronald Evans took the picture.

····→ Harrison Schmitt digs into the lunar surface with a specially designed tool in order to retrieve samples of the Moon's crust. His space suit's apparent dirtiness results from the composition of the lunar soil, which clings to surfaces.

reductions, Paine—imbued with the heady spirit of Apollo—pressed for an ambitious space program to succeed the conquest of the Moon. He persuaded the president to appoint a Space Task Group to give the administration a blueprint for the next decade. Paine won the panel's approval for a revolutionary new mode of travel, a spaceplane powered by rocketry and capable of being launched into orbit from conventional runways. By incorporating a "space tug" stage in the design, the group implicitly endorsed the goal of sending humans and materiel out into the solar system.

But Paine's expansive and improbable plans merely alienated the White House, which interpreted his dreams as empire-building. Many of Nixon's associates felt that by the early 1970s the space agency needed to forego its taste for adventure and instead concentrate on more attainable objectives. After only a year and a half in office, Paine resigned, leaving his deputy, George Low, in command. Low worked hard to moderate NASA's

appetites and to effect better relations between his agency and the White House.

Even as Paine left NASA and the Apollo program reached its zenith, a long series of robotic missions to the Moon and the planets started to vie with human spaceflight in the public imagination. These projects originated with the lunar satellites, probes, and landers enlisted to ascertain the composition of the Moon's surface before sending astronauts to walk on it. The overall success of these missions emboldened planners at NASA headquarters and at the Jet Propulsion Laboratory (JPL) in Pasadena, California, to conceive of a daring series of launches to shed light on the origins of the solar system.

The Jet Propulsion Lab began during World War II under the auspices of California Institute of Technology (Caltech) professor Theodore von Kármán, an internationally renowned authority on fluid mechanics and aeronautical engineering, a few of whose students asked him to add rocketry to the school's curriculum. Their experiments,

conducted under his supervision, resulted in a nascent center for rocket science that blossomed in the unlikely landscape of Arroyo Seco Canyon, located in the hills above Pasadena and the Caltech campus. The U.S. Army Ordnance and Artillery branches took notice of Kármán's group and began to contract for its research, a relationship that continued from the latter part of World War II until NASA acquired the laboratory from the Army on New Year's Day 1959. At the time of the transfer, JPL and NASA agreed to reorient the lab's portfolio from Army-sponsored rocketry to lunar and planetary exploration.

The JPL campus, led by no-nonsense director William H. Pickering, gradually assumed a prominence formerly reserved only for those parts of NASA involved in human spaceflight. However, during the months immediately following *Sputnik I*, Americans saw nothing but failure on their launch pads. Then, JPL's *Explorer I* entered the race. Engineers in Pasadena took only ninety days to integrate *Explorer*'s payload—a cosmic ray detector fashioned by Dr. James Van Allen—with a fourth-stage solid-propellant rocket motor. Technicians at Cape Canaveral, Florida,

hastily mounted this combination onto a Jupiter-C missile. Then, to the utter delight and relief of the American people, in January 1958 the Van Allen experiment became the first orbiting U.S. satellite. Not only that, but it discovered the immense radiation belts that encircle the Earth and that now bear Van Allen's name. But even before NASA and *Explorer 1*, Pickering (in a similar way to Hugh Dryden at the NACA) envisioned the importance of spaceflight and redirected JPL's energies from tactical Army missiles towards launch vehicle development. As a consequence, Pickering gradually weaned his engineers and scientists off pure research projects, and instead encouraged them to concentrate on operational rocketry (such as the successful solid-propellant Sergeant and the liquid-propellant Corporal systems).

Once in the NASA camp, JPL assumed a decisive role in the national space program. The lab started modestly with several lunar exploration programs conceived during the late 1950s. First came Ranger and Lunar Orbiter, but with the advent of Apollo and the need to know what awaited the astronauts on the Moon, they—along with

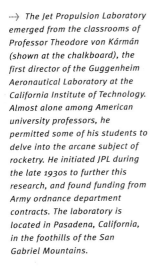

····⟩ *The Jet Propulsion Laboratory emerged from the classrooms of Professor Theodore von Kármán (shown at the chalkboard), the first director of the Guggenheim Aeronautical Laboratory at the California Institute of Technology. Almost alone among American university professors, he permitted some of his students to delve into the arcane subject of rocketry. He initiated JPL during the late 1930s to further this research, and found funding from Army ordnance department contracts. The laboratory is located in Pasadena, California, in the foothills of the San Gabriel Mountains.*

Surveyor—became servants of President Kennedy's great challenge (see Chapter 3). Collectively, these spacecraft developed a profile of the lunar landscape and environment.

Meanwhile, the JPL scientists pursued even more ambitious goals, based upon a daring series of planetary voyages. The first one bore the nautical name *Mariner*. This family of spacecraft actually originated during one of the lowest points in JPL history. Between August 1961 and October 1962, each of the first five Rangers—designed to land on the Moon's surface—failed to accomplish their mission. In contrast, the Soviets scored big successes in 1959 with the Moon flybys of the *Luna* spacecraft, and an even greater triumph when *Mars 1* flew by the Red Planet in June 1963. All through this period, members of Congress demanded answers and some talked of shutting down the Pasadena laboratory altogether. The reign of error continued in Mariner 1, a new project initiated in September 1961 with the objective of flying past the nearby planet Venus and gathering rudimentary data. Actually a Ranger spacecraft in thin disguise, it performed no better than its predecessors. Mounted aboard an Atlas–Agena rocket in July 1962, *Mariner 1* had to be destroyed due to a flawed trajectory during launch.

Shaken by debacle after debacle, Pickering and the JPL team took a big gamble. In the span of only thirty-six days after the loss of *Mariner 1*, they took another star-crossed Ranger spacecraft out of storage, modified it for the trip to Venus, called it *Mariner 2*, and launched it on August 27, 1962. It carried the identical suite of instruments as had *Mariner 1*. Just 20 pounds (9 kg) in weight, the probe had no capacity to photograph Venus's terrain, but it did carry sufficient equipment to evaluate the planet's atmosphere and climate. Somehow fated to brush time and time again with disaster, this spacecraft survived partly due to chance, and partly due to the resilience of the JPL ground crew. On launch, the Atlas–Agena booster went into a galloping roll that ended abruptly (and inexplicably) almost at the exact angle that ground controllers had desired from the beginning. Then the vehicle's Earth sensor fixed its gaze on the Moon instead of the Earth. Then a solar panel began shorting out repeatedly before finally refusing to work, forcing the mission to rely on the remaining one. Then the internal temperature of the spacecraft rose wildly. Its propellant tank gave readings of too much pressure. Near the approach to Venus, *Mariner*'s on-board computer failed to switch on the instruments, but fortunately a signal from JPL controllers awakened the spacecraft just as it made its pass.

Having survived blow after blow, in the end *Mariner 2* proved to be the champion JPL desperately needed, dissipating much of the political pressure on NASA, Pickering, and his team. Indeed, its journey constituted a

Thomas O. Paine

Thomas O. Paine had a short but influential career with NASA. Born in Berkeley, California in 1921 to Navy Commodore George T. Paine, he received an undergraduate degree in engineering from Brown University and a doctorate in physical metallurgy from Stanford in 1949. He spent the bulk of his career with General Electric, beginning as a research scientist in magnetic and composite materials, rising as a laboratory manager, as a center manager, and finally as manager of the firm's Center for Advanced Studies in Santa Barbara, California. Paine began with NASA as James Webb's Deputy Administrator in January 1968 and after Webb's removal by President Johnson, he assumed the roles of Acting Administrator and (from March 1969 to September 1970) NASA Administrator. His term witnessed the flowering of the Apollo program: the first seven Apollo missions with humans aboard occurred during his brief tenure. After resigning from the agency, he returned to General Electric where he became vice president. Paine remained a public advocate for a broad NASA mandate of exploration, and in 1985 he accepted an invitation from President Ronald Reagan to chair a National Commission on Space. The resulting report envisioned an expansive role for the United States in the cosmos, declaring the value of scientific research in space, as well as the exploitation of natural resources there, and human settlement on the Moon and Mars. Dr. Paine died in Los Angeles in May 1992.

⋯⋯⟩ *President Richard Nixon (left) announces the appointment of NASA Administrator Dr. Thomas O. Paine (center), as Vice President Spiro Agnew looks on. Among other problems, Paine faced the difficult task of succeeding James Webb, the man who guided Mercury, Gemini, and Apollo.*

William H. Pickering

A native of Wellington, New Zealand, William H. Pickering (born in 1910) grew up with his grandparents as a result of his mother's premature death and his father's on-going pharmaceutical research in the tropics. After matriculating his first year of college in New Zealand, he was advised by an uncle to apply for admission to the California Institute of Technology. Here, Pickering earned a Bachelor of Science degree in electrical engineering in 1932. He returned to New Zealand to find employment, but enrolled again at Caltech to pursue a master's and finally a doctorate in physics on the Pasadena campus. Pickering began teaching electrical engineering at Caltech upon graduating in 1936. Becoming increasingly at home in the United States, he decided to apply for American citizenship during this period and received it in 1941. Meanwhile, as the Jet Propulsion Laboratory formed under the guiding hand of Theodore von Kármán, Pickering became involved as a result of his main research interests, radio control and telemetry. By 1950, at the age of forty, Pickering had made the choice to stop teaching and join JPL full time. Here he led a research team developing the electronics for guided missiles and became project manager for the Corporal missile. His steadiness and determination impressed others, and in 1954 William Pickering assumed the role of director of the Jet Propulsion Laboratory, a position he retained until retirement in 1976. After many years in the wood fuel business, Dr. Pickering died in his home in La Cañada, California, at the age of ninety-three.

much-needed first for the United States during the early, embarrassing days of the space race. Its voyage constituted the first completely successful mission to another planet undertaken by any country. *Mariner 2* scanned Venus for forty-two minutes as it flew far above the surface on December 14, 1962, detecting a truly inhospitable world where a surface temperature of at least 797° F (425° C) seemed evenly distributed across the landscape. The spacecraft also discovered a thick cloud cover. Finally, still overheating, it swung past Venus and out into deep space, where NASA finally lost contact with it on January 3 the following year, 54.3 million miles (87.4 million km) from Earth and farther from its home planet than any object had ever traveled. An astonishing commentary on the healing power of success, the JPL turnaround cemented itself when President Kennedy invited James Webb, William Pickering, and several of the engineers from Pasadena to the Oval Office to congratulate them on their achievement.

Even before the triumph of *Mariner 2*, JPL engineers had begun to plan two far more ambitious journeys into the solar system, this time to Earth's brother planet, Mars. The Red Planet had provoked imaginations for generations. Like the Earth in some respects, its allure lay in the prospect that it might sustain life. Remarkably, the mission planners of *Mariners 3* and *4* envisioned trips that made the earlier one appear elementary, and received headquarters' approval even before *Mariner 2* had proven itself. Within two years (between November 1962 and November 1964) the JPL team had fabricated the *Mariner 3* and *4* probes, capable not just of sending pictures (in itself a leap beyond the earlier Mariners) but of detecting cosmic rays and dust, ionization, magnet fields, radiation, and other features of the Martian environment. Moreover,

←···· *American presidents liked to associate themselves with space successes. Here, the man who epitomized JPL, Dr. William Pickering (center), presents President Lyndon Johnson with photographs taken by the* Mariner *space probe. Pickering directed the laboratory for twenty-two years, from the days of the Sergeant and Corporal rockets to the crowning successes of the Grand Tour of the solar system. A few years earlier, President Kennedy receives a model of* Mariner 1 *in the White House.*

···→ Mariner 1, *designed and fabricated by the Jet Propulsion Laboratory, inaugurated a long series of planetary probes. In this photo, an Atlas–Agena rocket lifts* Mariner 1 *towards space in July 1962, but ground controllers had to terminate the flight shortly after launch.*

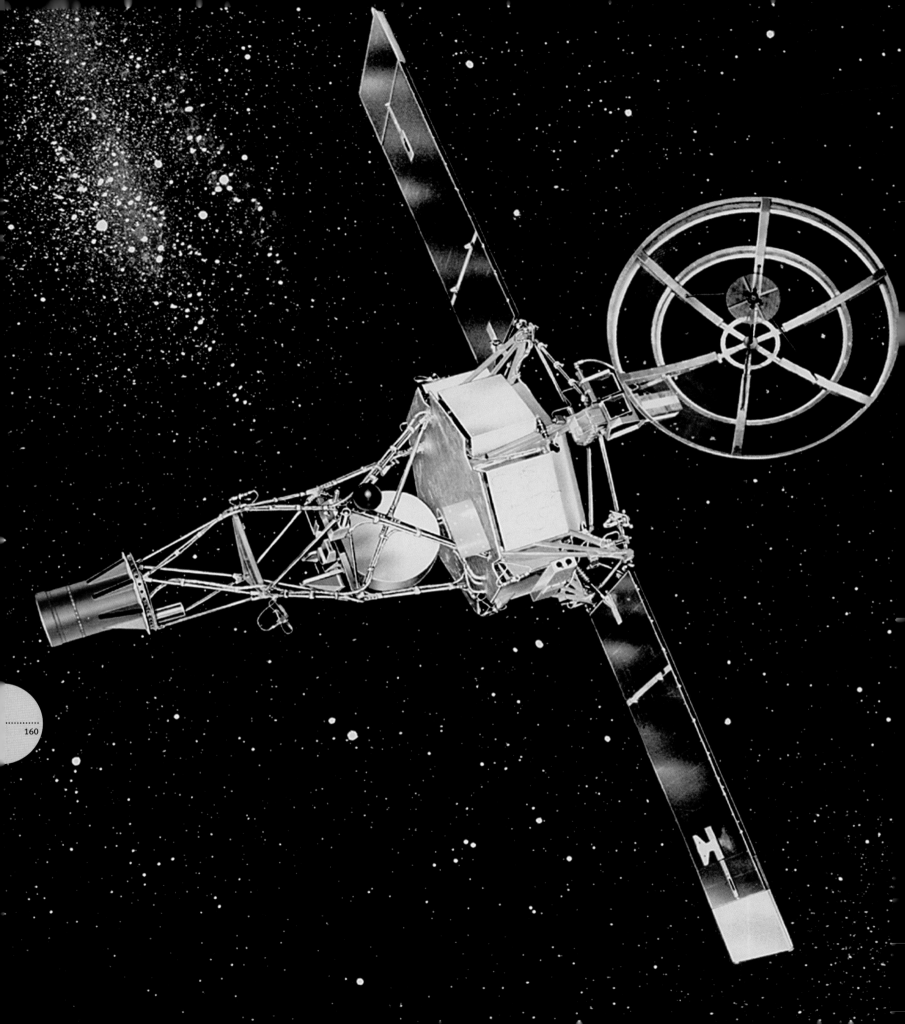

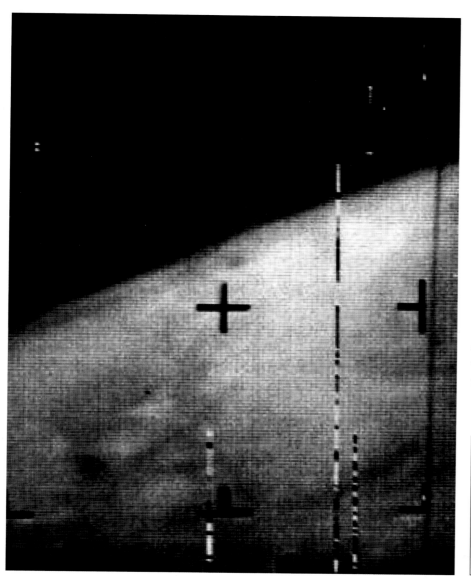

←···· Mariner 2 (pictured here) took off the month after Mariner 1 and proved to be a success. Four months after launch, it flew by Venus and took important readings of its surface temperature, as well as collecting data about solar winds.

Failures of the early planetary probes neither began nor ended with Mariner 1. Five consecutive Ranger spacecraft failed on their mission to land on the Moon. Mariner 3 had no better fate. But Mariner 4 became a triumph. It flew past Mars in July 1965 and sent home twenty-one photographs (one of which is shown here) and much valuable data on the planet's weather and atmosphere.

unlike *Mariners 1* and *2*, these two new spacecraft deviated from the earlier Ranger design, weighing over 570 pounds (260 kg) each, nearly 130 pounds (60 kg) more than the original Mariners.

As in the past, failures recurred. *Mariner 3* disappeared from ground control after its battery stopped functioning and its solar panels refused to open. But *Mariner 4* did much better. Launched on November 28, 1964, it approached Mars almost eight months later, on July 15, 1965. As it flew by at a distance of about 6000 miles (10,000 km), its camera captured twenty-one photographs that showed NASA's global audience a barren, Moon-like, cratered surface. Its instruments confirmed the bleakness of the Red Planet. Sensors recorded a thin atmosphere and bitter temperatures at the surface of minus 148° F (minus 100° C). There seemed small reason to hope for life. Disappointing as these results proved to those who

hoped for more, *Mariner 4*—like *Mariner 2* before it— bestowed another crowning success on the space agency.

Yet, the best of Mariner still lay ahead. Before the final missions, Mariners 5 to 7 further bolstered JPL's confidence. *Mariner 5* flew by Venus in 1967 and not only disproved a Soviet claim that one of their probes had landed there, but reliably detected a Venutian atmosphere ninety times the size of that of Earth. Two years later, *Mariners 6* and *7* went into space concurrently on a successful mission through the atmosphere of Mars. Because computing now allowed data processing to occur at over 16,000 bits per second (in another league compared to the 8.5 bits per second of *Mariner 4*), more advanced equipment went aloft this time, enabling JPL scientists to detect large concentrations of carbon dioxide and, most notably, water in the Martian atmosphere. After four consecutive successes, *Mariner 8* constituted yet another periodic—although gradually less frequent— loss. Launched in May 1971 to orbit Mars, the Centaur stage of the Atlas–Centaur rocket failed just after separation and the payload fell to Earth 900 miles (1500 km) from the launch site.

However, the last two Mariners (9 and 10) not only proved fruitful but actually inaugurated an age of planetary missions so successful that they vied with the triumphs of Apollo. They occurred largely during the long term of Administrator James C. Fletcher (April 1971 to May 1977), chosen by President Nixon to succeed Thomas O. Paine.

Fletcher's years represented a clear break with the form and substance of the human launches to the Moon. Unlike Paine, Fletcher accepted the bureaucratic imperatives of the time and abandoned the sweeping paradigm of Apollo. Instead, he embraced the tight-fisted realities of the day, as have all NASA administrators since. Fletcher set in motion many of the most enduring programs of the post-Apollo era, in effect becoming the architect of the American space program after the lunar missions. He also realigned the NASA centers in order to reduce duplication and preserve scientific and engineering talent in an era of persistent and deep cuts in the agency's funding. Most of all, Fletcher's era witnessed the decline of human missions (although the final three

161

HAVING SURVIVED BLOW AFTER BLOW, IN THE END *MARINER 2* PROVED TO BE THE CHAMPION JPL DESPERATELY NEEDED.

James C. Fletcher

Like Thomas Paine, James C. Fletcher began his career as a scientist. But Fletcher benefited from broader experiences than his predecessor. Born in Millburn, New Jersey, in June 1919, he attended Columbia University as an undergraduate and earned a doctorate in physics from the California Institute of Technology. Fletcher taught at Harvard and Princeton universities, joined Hughes Aircraft in 1948, conducted guided missile research at the Ramo-Wooldridge Corporation, became a co-founder of the Space Electronics Corporation, and served as Systems Vice President of Aerojet General Corporation. A Mormon, Dr. Fletcher assumed the presidency of the University of Utah in 1964. After President Nixon appointed him to be NASA Administrator in April 1971, Fletcher initiated a cluster of programs that had a profound influence on the agency's future: the Voyager space probes to distant planets, the Hubble Space Telescope, the Apollo–Soyuz partnership, and the Space Transportation System. He returned to NASA nine years after his first term when President Reagan persuaded him to lead the agency after the *Challenger* disaster. Fletcher oversaw the Shuttle's return to flight and retired again in April 1989. Dr. Fletcher died at his home in the Washington, D.C., suburbs in December 1991.

Apollo flights did happen under his watch) and the zenith of planetary probes, sometimes referred to as the golden age of planetary exploration.

At the start this so-called golden age, *Mariner 9* (launched on May 30, 1971) represented a quantum leap in spacecraft complexity. With a weight four times that of *Mariners 3* and *4* (at nearly 2200 pounds/1000 kg), wide and octagonal in shape, nearly 23 feet (7 m) across with its solar panels extended, this spacecraft pivoted on a moveable platform beneath the main mass of the vehicle, on which technicians installed its instruments. After a passage of nearly six months, on November 14, 1971, *Mariner 9*'s main engine awakened over Mars, propelling it into the first orbit ever achieved by an artificial body around another planet. The probe was packed with an imaging system, an ultraviolet spectrometer, an infrared spectrometer, and an infrared radiometer, and its mission planners hoped to keep it circling long enough to take pictures of 70% of the surface.

At first, luck failed them. Mars at this period happened to be orbiting very close to the Sun, resulting in Martian winds up to 400 miles (644 km) per hour. A yellow dust cloud soon covered the planet, obscuring *Mariner*'s view—except for a white cap on the South Pole—until January. But when clear pictures began to arrive, scientists and the general public marveled.

The earlier Mariners left the impression of a dead planet, like the Moon. But viewing different parts of the

←⋯ *James C. Fletcher succeeded Thomas Paine as NASA Administrator. Fletcher accepted President Nixon's insistence on a more frugal U.S. space program and managed NASA accordingly. He served twice as Administrator (about nine years in all) and presided over a succession of highly successful planetary probes, as well as over the development of the Space Shuttle.*

⋯⟩ ⋯⟩ ⋯⟩ *The Mariner probes continued to have successes and failures. Mariner 5 left Kennedy Launch Complex 12 in December 1967 aboard an Atlas–Agena and on its trip past Venus made important observations about its atmosphere. Mariner 8, illustrated here as technicians at Kennedy install a solar panel, did less well. After launch, the Centaur second stage failed, ending the mission.*

MARINER 5 NOT ONLY DISPROVED A SOVIET CLAIM THAT ONE OF THEIR PROBES HAD LANDED ON VENUS, BUT RELIABLY DETECTED A VENUTIAN ATMOSPHERE NINETY TIMES THE SIZE OF THAT OF EARTH.

Mariner 9's mission to orbit Mars inaugurated a watershed period in American spaceflight, in which a series of extraordinarily successful planetary probes unlocked many of the mysteries of the solar system. Opposite, a technician examines the Mariner 9 prior to encapsulation. Once completed and mated to the Atlas–Centaur rocket (above left), the spacecraft underwent radio frequency interference tests. Finally, on May 30, 1971, the stack rose from Kennedy Launch Complex 36B.

terrain over and over again presented a startlingly different picture. Earlier images of dark spots proved to be gigantic volcanoes, the biggest some 70,000 feet (21,000 m) high with immense lava flows adjacent. Near the largest volcano, a massive canyon as wide as the United States and 4 miles (6.4 km) deep scarred the Martian landscape. Scientists who observed this groove in Mars closely also noticed that tributaries flowed into it, and that streams flowed downstream and out, suggesting flooding at some distant time. Clearly, water existed not just in the planet's atmosphere but at its poles, and had previously existed on other parts (presumably, when the volcanoes melted permafrost and caused flooding for a time). Therefore, the prospect of life of some kind seemed not at all far fetched. When *Mariner 9* at last closed down in October 1972, it had mapped 85% of the planet and had taken 7329 pictures at a resolution of about half a mile to a mile, leaving no doubt about what had been seen.

----> ----> *The* Mariner 10 *mission represented the first time in planetary exploration that a probe had flown past two planets, in this instance Venus and Mercury. Prior to the launch, the Atlas–Centaur combination underwent tankage tests. Then, in November 1973, the lift-off occurred, relying on this smaller, less powerful launch vehicle rather than the more potent Titan booster because of another first: the alignment of the Sun and planets enabled the flight to be assisted by solar gravity.*

One more daring Mariner mission remained. Leaders at JPL decided to direct their attention to Mercury, not yet explored, the smallest of the planets, and the closest to the Sun. Yet, at the time when their proposal went to Headquarters (1969), Apollo continued to absorb the bulk of the NASA budget. Moreover, a mission to a presumably "dead" celestial body offered little appeal to members of Congress. But members of the *Mariner 10* team found a cogent "selling point." The year 1973—*Mariner 10*'s planned launch period—offered an opportune time to visit both Mercury and Venus. During that year, the alignment of the Sun and planets enabled the spacecraft to travel by gravity assistance, the first use of this technique ever attempted. During normal years, the mission would have required the thrust of the big and costly Titan IIIC–Centaur launch combination; using the extra force of gravity available in 1973, the smaller, cheaper Atlas–Centaur had enough power. Moreover, including Venus in the mission heightened the appeal of a planet in some ways similar to Earth. But many technical difficulties existed (such as the Sun's powerful gravitational force, which added to the spacecraft's speed), as did programmatic problems related to the tight budgetary ceiling of $98,000,000 imposed by JPL director Pickering.

In the end, the *Mariner 10* mission—launched on November 3, 1973—nonetheless proved to be memorable. It became the first spacecraft to visit two planets, the first to take advantage of gravity assisted techniques to alter its flight path, and the first to return to a planet after an initial encounter. *Mariner 10* approached Venus on February 5, 1974, and collected more than 4100 images as it passed. Then, propelled by the Venutian gravitational force, the spacecraft headed to Mercury. It made three passes. In March 1974 it flew as close as 437 miles (703 km) and found a barren, Moon-like topography. The following September, after flying around the Sun, it returned for a second look. Finally, the last of the Mariners made one more loop around the Sun, and in March 1975 came as close as 191 miles (307 km) from Mercury.

Well before the end of Mariner—in fact, as early as the mid-1960s—JPL embarked on an even more audacious project. NASA scientists and engineers wanted to expand the successes of Mariner, in part motivated by the

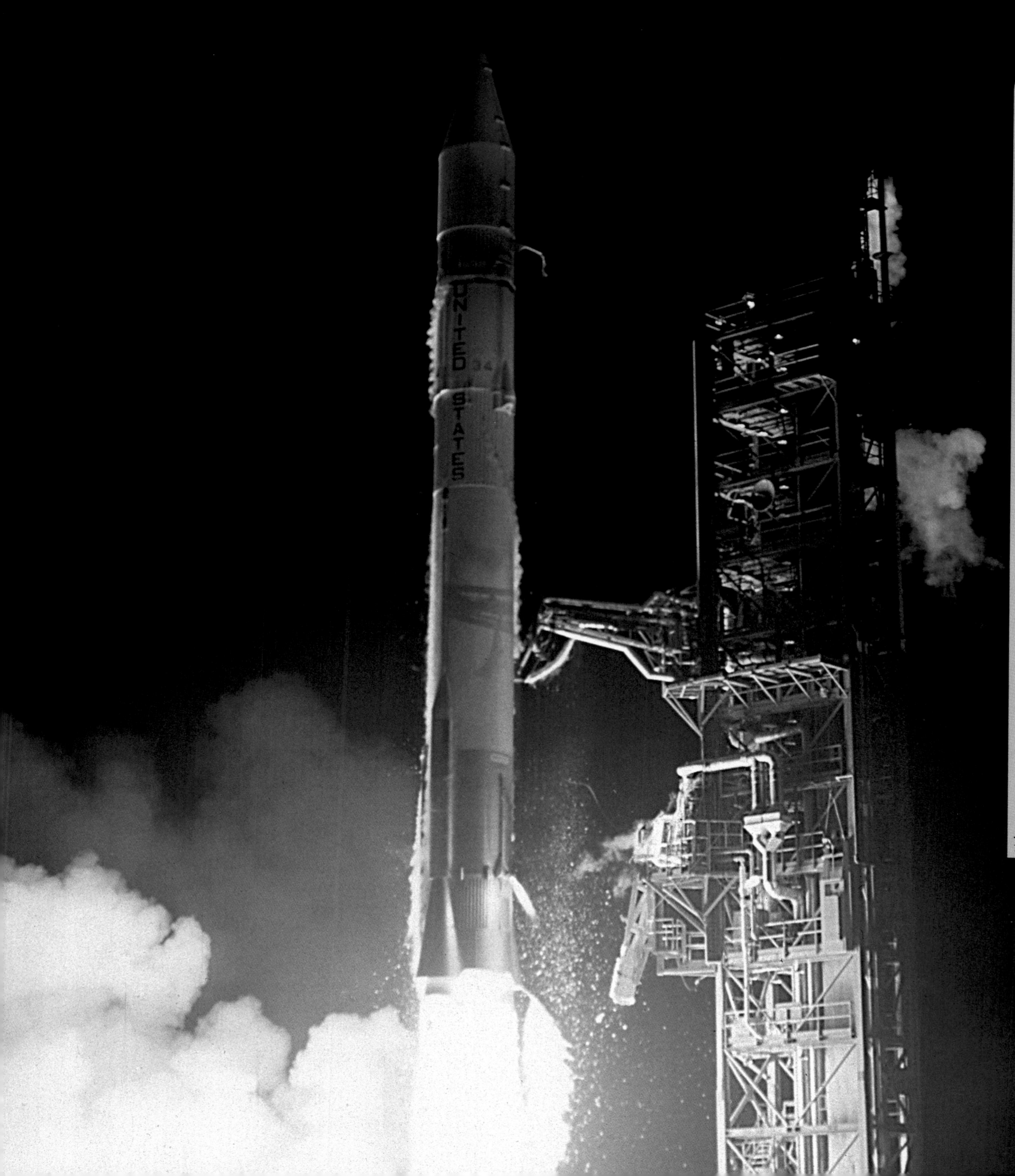
UNITED STATES

↑ *Not content to pass by Mars or to orbit it, NASA researchers at Langley Research Center and elsewhere planned during the late 1960s to develop a probe to land softly on the Red Planet. Known as* Viking, *it featured a large disk called an aeroshell (pictured here) to protect the lander during entry into the Martian atmosphere.*

frequent (but unsuccessful) attempts of the Soviet Union to enter the Mars sweepstakes, and in part due to the irresistible momentum of the Apollo program. Above all, it was the desire to penetrate the conundrum of life on Mars that inspired the missions, and NASA headquarters called the new program Voyager Mars. The JPL scientists, as well as researchers from the Ames and Langley research centers, conceived of an expensive, big, and bold project. But the Voyager orbiter–lander–rover combination grew large and heavy with requirements, so much so that the spacecraft required a *Saturn I* rocket for launch. When the *Saturn I* went out of production, leaving the mighty

Saturn V as the only launch vehicle capable of hoisting the spacecraft, scientists took the opportunity of its immense lifting capacity to add still more to Voyager Mars. Their ambitions ultimately resulted in a project projected to cost about $5,000,000,000, far too much in a time when the Apollo funding profile was in nosedive. In late 1967, Congress declined to back Voyager Mars.

Still, the idea of investigating life on Mars remained compelling. Inside of NASA, Titan Mars rose on the ashes of Voyager Mars. All through this period, JPL and Langley teams continued to study, and eventually compete for, the project. In the end, officials at NASA headquarters

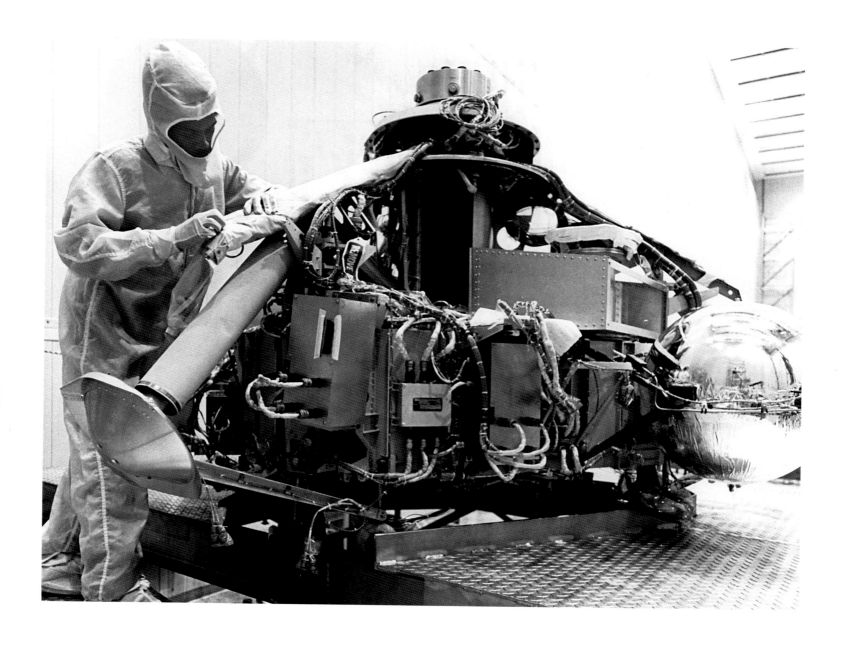

↑ *Powered by two nuclear generators, the Viking lander (illustrated in a cleanroom at Martin Marietta Aerospace) included stereo cameras, a weather station, and biological instrumentation to detect life.*

decided that Langley showed better promise to manage the mission, and planning began in 1968. Renamed *Viking* by James Webb's successor Thomas Paine, its proponents appeared before Congress and pledged a soft-lander spacecraft for $750,000,000. Budget cuts in the Nixon administration postponed *Viking*'s development for two years, delaying the first launch until 1975, the next year in which the planetary mechanics at least partially favored a trip to Mars.

The postponement proved both good and bad. It gave Langley's scientists and engineers extra much-needed time to cope with the complexities of a Mars landing; but

it also resulted in more expense due to inflation and to the less than optimal celestial alignments. During this period, the budget rose to $830,000,000, a fact grudgingly accepted by Congress. Acting as prime contractor, Langley chose its partners: Martin Marietta became responsible for fabricating the landers, JPL won the assignment to build the orbiter (in-house, applying the wealth of Mariner experience), and Lewis Research Center (collaborating with Martin Marietta) took charge of Titan III–Centaur launch vehicle development.

The JPL team decided to take the course of safety and design the orbiter in the mold of *Mariner 9*, only bigger

[Left] *Because of* Viking's *size and weight, the Atlas–Centaur combination used for Mariner had to be abandoned in favor of the Titan II launch vehicle used for the Gemini flights. To test this stack (in which Centaur ignited after the second stage of Titan burned away), a rehearsal flight of the Viking system occurred in January 1974 (shown here, prior to launch).*

[Opposite] *Two views of the dramatic launch of the much anticipated Viking 1, August 20, 1975, from Complex 41 at Kennedy Space Center. Viking traveled for 310 days before it went into orbit around Mars.*

TAN/CENTAUR COMPLEX 41

[Above and top] *During the American Bicentennial month, Viking 1's orbiter and lander began to send back images. As the orbiter scanned the planet looking for patterns relating to climate and topography, the lander showed researchers their first, close-up look at the surface. The initial photograph depicted the ground next to the lander, showing the disturbance associated with the touchdown. Three days later, NASA received the first panoramic shot of Mars, showing terrain and sky.*

[Right] *Even better images followed the initial two. On July 24, Viking 1 transmitted in highly vivid color a Martian landscape complete with boulders, a broad area of sky, and some of the lander itself.*

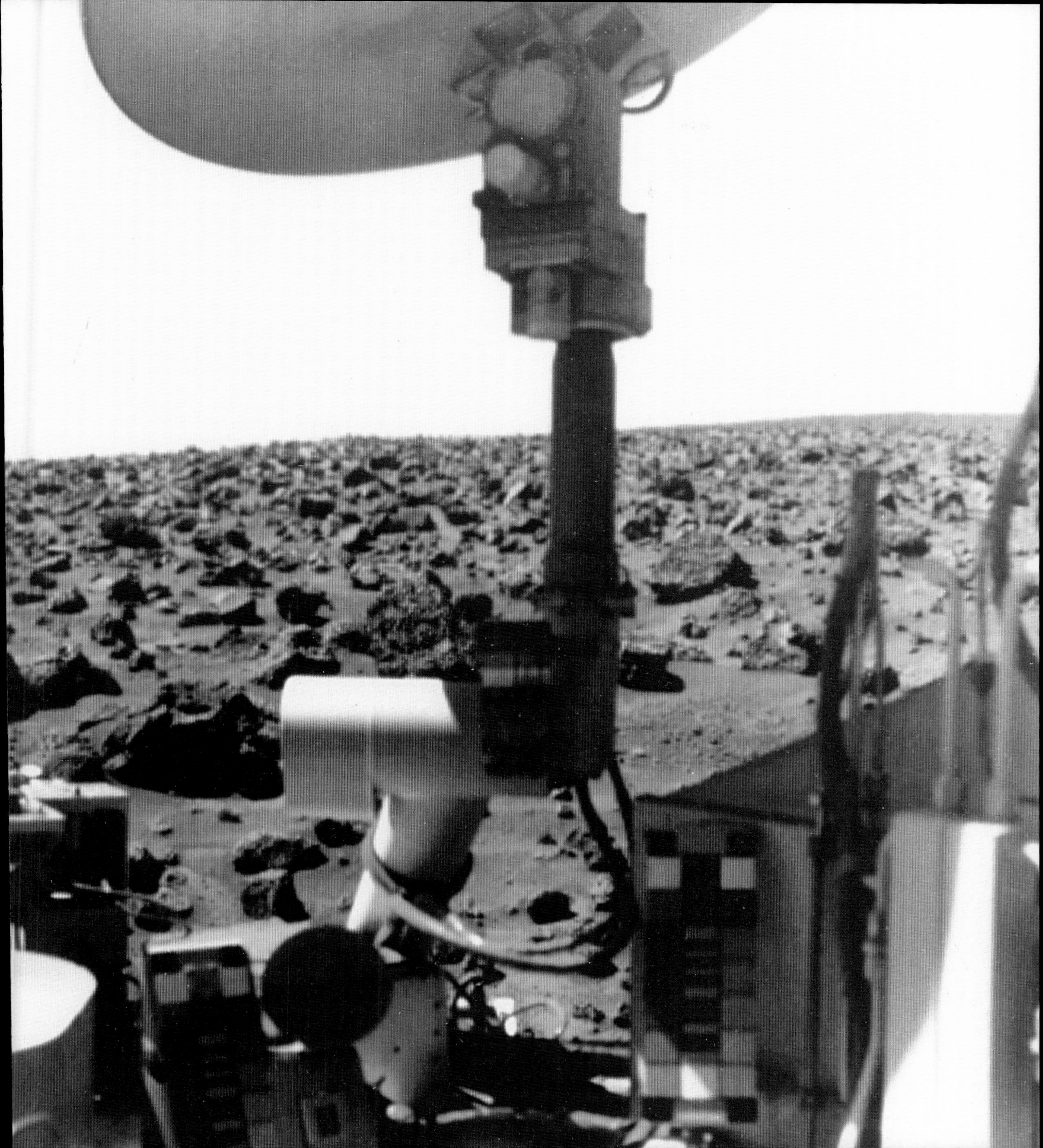

and with advanced imaging and communications. The lander, on the other hand, remained fraught with difficulties because of the technical demands, some self-imposed, some not. For instance, Langley's Viking program manager, James Martin, insisted not just on twin cameras for stereo depth perception but on full color as well. In addition, the lander's chromatograph-spectrometer, essential for sensing organic substances, represented a breakthrough in the art. One of the senior officials in the program recalled feeling almost overwhelmed by the technical complexities:

It was soon apparent that we had bitten off more than we could chew. On Earth these four [biological] experiments would require a large laboratory crammed with equipment; on the Viking Lander we were allotting just one cubic foot, little more than the size of a gallon milk carton. Into that cube must go multiple ovens, many ampules of nutrients, radioactive-gas bottles, a solar simulator, many dozens of valves, and a computer containing twenty thousand transistors to adaptively program successive experiment procedures should the Lander lose its uplink commands from Earth.[2]

Likewise, the lander's own hardware posed problems. For example, the memory of the vehicle's central computer resided in matrices of plated wires just 0.002 inches

(0.05 mm) thick, produced under a new and complicated process. The contractor—the Honeywell Corporation—delivered the computer so late that it had to be installed at Cape Kennedy and tested just before final launch preparations. Under such pressures, manufacturers missed deadlines, the government felt obliged to improve contract incentives, and costs rose. Only because of some budgetary maneuvering by NASA's accountants (who fashioned a pool of reserve funds from unrelated sources to cover Viking's deficits) did a number of space science projects manage to avoid cutbacks or cancellations.

At last, the time to launch *Viking 1* arrived. It lifted off from Kennedy Space Center on August 20, 1975 and entered Martian orbit nearly ten months later, on June 19, 1976.

Once *Viking 1* achieved orbit around Mars, images of the surface showed scientists that the initial landing zone appeared too rough, so they redirected the lander to a safer point. After about a month circling the planet, the lander (just 2156 pounds/978 kg) separated from the orbiter (5157 pounds/2339 kg) and after a series of complex maneuvers, landed safely. Despite the realization of this engineering marvel (the first such feat accomplished on another planet), Viking's researchers experienced both elation and dejection. Its color

←⋯↑ *After a routine launch from Kennedy (Pad 41), Viking 2 entered Mars's atmosphere in August 1976, also landing successfully, some 4600 miles (7400 km) from Viking 1. This panorama, sent home in August 1976, shows a landscape strewn with boulders and a salmon-colored sky (which scientists attributed to red dust particles suspended in the atmosphere).*

Bruce Murray

During the period of *Viking 1*'s race to Mars, many at JPL held their breath, not only because of the mission but also because the laboratory was undergoing a decisive change in leadership. After a remarkably long and fruitful twenty-two years of service, Dr. William Pickering stepped down, to be succeeded in April 1976 by Dr. Bruce Murray. A geologist (in contrast to Pickering's background in physics), Murray (born in November 1931) grew up in New York City. An energetic and determined personality, he cut his teeth at JPL on Mariner as co-investigator in the television experiments in numbers 4, 6, 7, and 9. Not only did he serve as team lead of the *Mariner 10* television experiment but his active role in the overall mission earned him the admiration of his fellow researchers. Like his predecessor, Murray began his career as a professor at the California Institute of Technology (Caltech), but unlike Pickering, he took his education elsewhere, namely the Massachusetts Institute of Technology. Murray remained on the Caltech rolls from 1960 to 2001 (when he became emeritus), even as he led JPL from 1976 to 1982. He guided JPL during some of its most resounding successes, such as the two Viking reconnaissance trips to Mars, and the so-called "Grand Tour" missions to the outer planets (Jupiter, Saturn, Uranus, and Neptune). Murray remained an active collaborator in planetary explorations, including the Mars Global Surveyor, the Mars Polar Lander, and the Mars Microprobes. With Carl Sagan and Louis Freedman, he founded the Planetary Society and served as its president.

↑ *Because of the longevity of the Pioneer program, the probes themselves and the supporting equipment evolved greatly. Pioneer 1 went into space (but failed to reach the Moon) on an intermediate ballistic missile known as Thor, equipped with the Able second stage.*

photographs proved to be astounding. It also transmitted weather updates (lows of minus 123° F/minus 86° C at dawn to minus 27° F/minus 33° C in the afternoon) until February 1983. But the hunt for even the simplest life forms remained unfulfilled. Although the robot arm on the lander succeeded in scooping a soil sample into its biological laboratory, and some of the resulting data hinted at the possibility of life, no organic compounds came to light.

A month after *Viking 1* left on its mission, *Viking 2* set sail for the Red Planet on September 9, 1975. It entered orbit about eleven months after launch (August 7, 1976). Mission controllers soon realized that the planned touchdown site of the lander had to be changed, as in *Viking 1*. So they directed it to a zone near the edge of Mars's polar ice cap. They hoped that the evidence of life

might be more conclusive in that area. Once again, the lander touched down perfectly. But, as before, the scoop tests also returned indefinite results. Meanwhile, the orbiter revolved around Mars repeatedly, photographing (with the contributions of *Viking 1*) 97% of the planet and transmitting over 51,500 images. Thus, the knowledge about Mars increased immensely, but the question of Martian life remained tantalizingly unresolved.

Still, the golden age of planetary exploration had not yet passed. The versatile family of planetary vehicles known as Pioneer—nine of them in all—flew into the heavens over an incredibly long period of time, from 1958 to 1973. Originally one of the earliest Moon probes, the Pioneer series later assumed a unique role as observers not so much of a particular celestial body but of the vast spaces between the planets. The first, a tiny (13 pound/

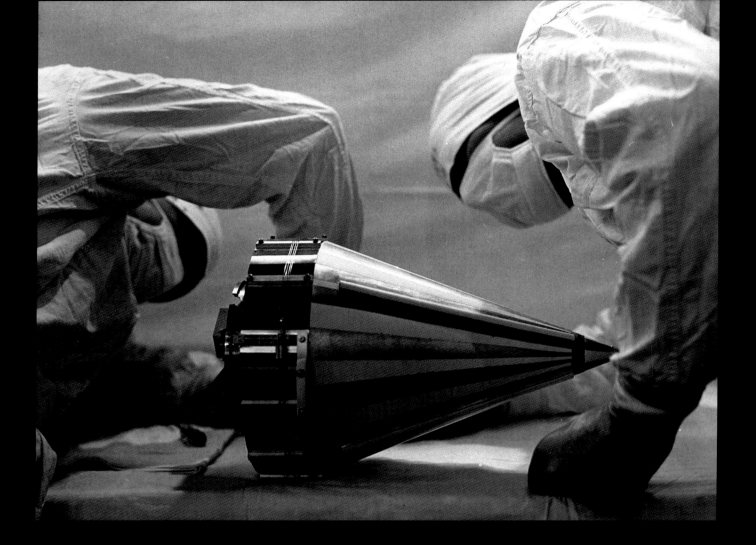

[Opposite, top] Pioneer 3, launched in December 1958 aboard a four-stage Jupiter missile (called Juno II), undergoes inspection by cleanroom technicians before take-off. Designed to determine the intensity of the Van Allen radiation belt, among other things, the mission ended when the Juno failed to reach sufficient escape velocity.

[Opposite, bottom row] *The* Juno II *served more faithfully on the launch of* Pioneer 4. *The booster sent the probe out of the atmosphere and it became the first U.S. spacecraft to orbit the Sun, although it missed the other goal of passing close enough to the Moon to take photographs. From mission control, Wernher von Braun (top right of photograph, second from right) watched the launch.*

[Left and below] *Observers watch as* Pioneer 6 *is mated with the Delta third stage launch vehicle. This time, project planners selected the Thor booster, which sent* Pioneer 6 *on its mission in December 1965 — to orbit the Sun and collect data.*

THE PIONEERS UNDERWENT A METAMORPHOSIS.
DURING THE EARLY 1960S, NASA HEADQUARTERS
DECIDED TO USE THEM AS LONG-LIVED OBSERVERS OF
SOLAR WIND AND OTHER PLANETARY PHENOMENA.

6 kg) vehicle known as *Pioneer 3*, launched from an Army *Juno II* rocket in December 1958. It had been conceived by JPL and the Air Force Space Technology Laboratories, and from it scientists hoped to gain data from a lunar flyby. But the spacecraft failed to gain escape velocity and burned up in the Earth's atmosphere. About three months later the team tried again with *Pioneer 4*. This time it worked—to a degree. Although it became the first American probe to escape the Earth's gravity, it passed too far from the Moon for its sensing equipment to take pictures. A much different Pioneer flew in March 1960. Over 95 pounds (43 kg) in weight, *Pioneer 5* (a partnership of the NASA Goddard Space Flight Center and the USAF Ballistic Missile Division) left the Earth aboard a Thor–Able IV combination that lofted it on a direct path to orbit the Sun. It subsequently entered a heliocentric path around our familiar star, passing between Earth and Venus and discovering magnetic fields between the planets.

Afterwards, the Pioneers underwent a metamorphosis. During the early 1960s, NASA headquarters decided to use them as long-lived observers of solar wind and other planetary phenomena. NASA Ames Research Center led the project, and contracting with TRW, Inc., launched *Pioneer 6* (aboard a Thor–Delta E stack) in December 1965. On its long journey around the Sun, it relayed data about the solar atmosphere, traced the tail of Comet Kohoutek, and most importantly, served as a weather station that predicted solar storms for a host of government and business customers whose electronic systems depended on such information. Initially expected to survive for six months, *Pioneer 6*'s main transmitter failed in December 1996—no fewer than thirty-one years after launch. The space agency has switched on the spacecraft intermittently. Three brother probes followed, which together served as a constellation designed to measure the effects of solar storms on Earth: *Pioneer 7*, launched in August 1965; *Pioneer 8*, launched in December 1967; and *Pioneer 9*, launched in November 1968. *Pioneers 7*, *8* and *9* served much the same functions as *Pioneer 6*, although the last one placed in orbit an Earth satellite that tested ground-to-space communications for the Apollo program.

←···· *Kennedy Space Center technicians work on* Pioneer 10 *after delivery by its manufacturer, TRW Systems. Researchers hoped for seven years of service from* Pioneer 10; *it actually continued for twenty-four years.*

····→ *The Pioneer program operated during the 1950s, 1960s, and 1970s.* Pioneer 10, *launched in March 1972, traveled to Jupiter for a close flyby, and continued toward the outer solar system and beyond. Seen here in the space simulation chamber at TRW Systems, the spacecraft underwent trials in many of the harsh conditions encountered in space.*

····→ ····→ *The Pioneer program finally ended with* Pioneer 11. *Launched by a time-honored Atlas–Centaur combination from Kennedy's Complex 36 in April 1973, it flew by Jupiter, then Saturn in 1979, and left the solar system in the opposite direction to* Pioneer 10.

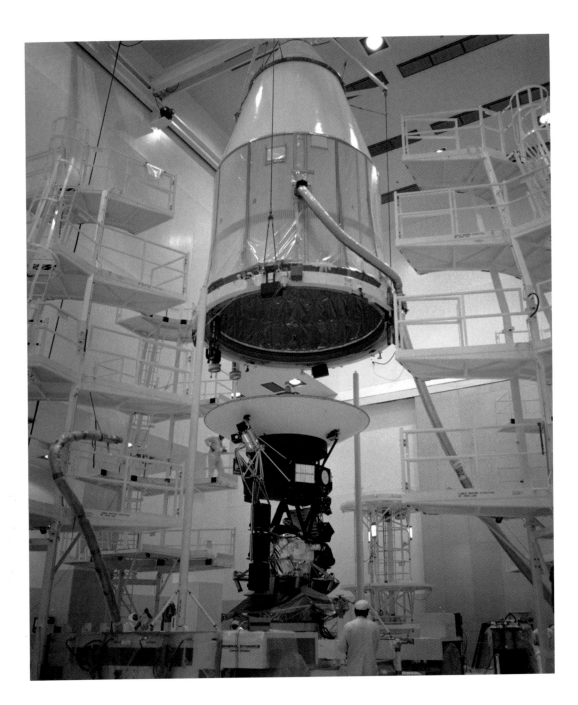

The "Grand Tour" envisioned by JPL entailed a flyby of the outer planets, occasioned by their unusually close alignment to one another. Here, one of the two spacecraft designated for the project—Voyager 2—undergoes encapsulation at Kennedy Space Center in preparation for launch.

Standing on Launch Complex 41, the encapsulated Voyager 2 and its launch vehicle are readied for flight. The booster consisted of a Titan III—Centaur 7 combination, a big rocket necessary to lift the 1800 pound (816 kg) spacecraft. The probe eventually flew past Jupiter, Saturn, Uranus, and Neptune.

The final two Pioneers carried the legacy of the last four vehicles—long-term, long-distance interplanetary probes—to a new level of exploration. As early as 1967, the planetary fraternity in NASA began to dream of a mission to Jupiter, the giant of the planets with a mass twice that of the rest combined. Scientists and engineers at the Ames Research Center and the Goddard Space Flight Center vied for the project. Ames won when headquarters officials decided that the California center and its industry partner TRW, which had succeeded so well on *Pioneers 6* to *9*, deserved to manage *Pioneers 10* and *11*. The technologists faced a tall hurdle in this instance. The solar energy coincident to Jupiter, located so far from the Sun, could not produce sufficient power to animate the spacecraft's many and complex instruments. Thus, unlike all other probes and satellites until this time, the instruments and controls aboard the spacecraft operated not by conventional solar panels but by all-nuclear electrical power generated by equipment supplied (without cost to NASA) by the Atomic Energy Commission.

The first of the Pioneers weighed but 13 pounds (6 kg); the last, 569 pounds (259 kg). The final ones required the best of present-day technology to send them to their distant targets (traveling at 32,188 miles/51,800 km per

plaque illustrating a man and a woman, the location of Earth, and when the machine left its home planet.

Its sister ship, *Pioneer 11*, rocketed from Earth in April 1973. It journeyed first to Jupiter and, flying faster (106,000 miles/171,000 km per hour) than any mechanical object to that time, came three times closer to the planet than its predecessor. Using Jovian gravity, the spacecraft changed course across the solar system for a look at Saturn. In September 1979 it crossed the planet's ring plane, flew just 12,987 miles (20,900 km) from the surface, took pictures of the relatively flat landscape, determined that it consisted mainly of liquid hydrogen, and detected average surface temperatures of minus 290° F (minus 180°C). Then *Pioneer 11* oriented itself in the opposite direction to *Pioneer 10*, left the solar system passing Neptune in February 1990, and continued on its path into the cosmos bearing the same plaque showing the spacecraft's origins. NASA ended contact with the last of the Pioneers in September 1995.

These missions, like the rest of American space exploration, owed their existence in part to public enthusiasm and to congressional support, but perhaps even more to presidential initiatives. Just as John F. Kennedy dominated the formative first decade of spaceflight, Richard M. Nixon left a powerful imprint on space travel during and after the 1970s. Nixon favored ventures that balanced robotic missions with human ones, but declared in 1970 that a "Grand Tour" of the outer planets represented the capstone of his policy. With the last of the Pioneers, this goal took precedence.

The stagnant economy that shadowed his years in office profoundly affected Nixon's space policy. Although he profited politically from the Apollo and planetary successes charted during Kennedy's term, he also felt that the space agency had proved its supremacy in the heavens. He therefore wanted NASA to retrench, not just because of federal deficits but also due to polls that showed that 64% of Americans felt that the agency's $4,000,000,000 budget needed to be trimmed. Accordingly, he told Administrator Thomas Paine to reduce to $3,000,000,000 annually. He also wanted NASA to retreat from headline-grabbing events and programs and to assume the role of a more mature institution.

Nixon did so in the face of findings by his own Space Task Group, which in 1969 endorsed a bold post-Apollo program that included a space station that would serve as a way station to the Moon and Mars. The president responded the following March, and not favorably. He conceded the importance of space discovery, but insisted it be pursued at lower payload launch costs, with greater international cooperation, and in a less visionary, more practical manner. Nixon's support of the automated tour of the planets and of the solar system as a whole

↑ *Despite their numerical order,* Voyager 2 *actually left Kennedy before its twin,* Voyager 1. *In this picture, a technician prepares to equip* Voyager 1 *with a gold-plated record containing typical sounds heard on Earth, both natural and artificial.* Voyager 2 *carried the same item, designed to inform whoever might intercept the Voyagers on their long journeys.*

hour, a record initial speed for launched objects). Sent in March 1972 on a trajectory to leave the solar system, *Pioneer 10* faced its first challenge at the asteroid belt between the orbits of Mars and Jupiter. To the elation of everyone—not least the project managers—in February 1973 it left this danger zone with some hits, but without significant damage. This feat gave hope, because until this point, no probe had flown farther than Mars. By November 1973 *Pioneer 10* had begun to transmit the first of 300 images of Jupiter, and the following month it approached the mighty planet at its closest point (more than 80,998 miles/130,354 km). The spacecraft flew on, crossing Saturn's orbit in February 1976 and speeding past that of Neptune in June 1983, at which time it became the first object fashioned by humans to exit the solar system. NASA ended contact in March 1997, although the spacecraft continues to fly into deep space and carries (for any creatures that may intercept it) an aluminum reflected these more conservative leanings. His cornerstones—planetary exploration, lower launch cost per pound of payload, and greater international collaboration—had a profound influence on the American space program, and continue to do so to this day.

President Nixon abandoned his office when it became apparent that he had ordered and then denied the wiretapping and a break-in of the Democratic National Committee offices in Washington, D.C. He became the first U.S. president to resign in office, turning over the reins of government to Vice President Gerald R. Ford on August 8, 1974. The former president lived for almost twenty years after his resignation, and to some extent succeeded in rehabilitating his reputation. He died on April 22, 1994 in New York City.

Luckily, Nixon's space policy coincided with the alignment of nature. Scientists at the Jet Propulsion Laboratory realized that between 1976 and 1979 the relative position of the outer planets enabled spacecraft to advance by gravity assist past Jupiter and on to distant bodies in the solar system. Called the "Grand Tour" by JPL officials, it enjoyed the personal backing of Administrator Paine, the conceptual support of President Nixon, and at length the nod from Congress and the Office of Management and Budget. Although the idea did encounter opposition from competing projects at other NASA centers, eventually the JPL vision took root: a voyage to the far reaches of the solar system pursued by relatively inexpensive spacecraft that represented advanced versions of the Mariner design. Moreover, to ensure a sustained period of intense activity at the lab, JPL leaders decided to conceive and fabricate the Grand

Richard M. Nixon

A man of solid intelligence, Richard Milhous Nixon became one of the most controversial figures in American political history. Born in Yorba Linda, California, on January 9, 1913, his father ran a gasoline station and his mother Hannah—a pious Quaker—raised five sons. He received a scholarship to Duke University Law School, where he graduated third in the class of 1937. Upon discharge from the Navy, Nixon ran for Congress in 1946. A hard campaigner not afraid of invective, Nixon won easily. He then won election to the U.S. Senate from his native state in a campaign during which he questioned the patriotism of his opponent. During the following two years, Nixon came to the attention of presidential candidate Dwight D. Eisenhower. He ran as Eisenhower's vice president and won election twice. His party turned to him in the 1960 presidential contest against Senator John F. Kennedy. In the end, Kennedy won by a scant 113,000 votes. Against long odds, Nixon revived his political fortunes and ran for the presidency again in 1968, defeating Vice President Hubert Humphrey by only 500,000 votes among some 73 million cast. Nixon won many successes as president, most notably improved relations with the Soviet Union, successful diplomatic overtures to China, and a ceasefire in the Vietnam War. But he also suffered politically and personally from opposition to the war, struggled to keep the federal budget from ballooning, and ultimately resigned from office after the Watergate scandal. He died in 1994.

←⋯ Lifting off in August 1977, Voyager 2 eventually flew out of the solar system, at which point NASA refocused the project, calling it the Voyager Interstellar Mission, on which it continues to travel.

⋯→ President Richard M. Nixon (far right) greets Apollo 11 astronauts (left to right) Buzz Aldrin, Michael Collins, and Neil Armstrong on the White House lawn in 1969. The crowd behind them had gathered to celebrate the astronauts' return after an exhausting forty-five-day Presidential Goodwill Tour to twenty-four countries.

⟵···· Voyager 1 *left Earth the month after* Voyager 2, *but it arrived at the edge of the solar system first because of the more direct path it took to Jupiter and Saturn. It also lifted off on the back of the Titan 3–Centaur 7 pairing.*

The Jet Propulsion Laboratory assembled an impressive montage showing images taken by the Voyagers as they sped by four planets (right to left): Jupiter, Saturn, Uranus, and Neptune.

An even more impressive montage, showing images assembled from all of the JPL-sponsored missions to the other planets. From top to bottom, they include: Mercury, Venus, Earth (with Moon), Mars, Jupiter, Saturn, Uranus, and Neptune. The smaller planets are roughly in scale to each other, as are the bigger planets to each other.

Tour vehicles on campus in Pasadena, rather than outsource them to private industry.

The pair, called *Voyager 1* and *2*, arrived on schedule at Kennedy Space Center in summer 1977. Mounted atop a Titan IIIE–Centaur stack, *Voyager 2* flew into space in August 1977 on its way to flybys of Jupiter, Saturn, Uranus, and Neptune. It reached Jupiter in April 1979 and transmitted spectacular photos of the planet and its moons Amalthea, Io, Callisto, Europa, and Ganymede, as well as movies showing the circulation pattern of the Jovian atmosphere. *Voyager 2* then traveled to Saturn where its moons Hyperion, Enceladus, Tethys, and Phoebe also revealed themselves to the prying eye of the camera, as did the planet's stupendous rings. In January 1986 the spacecraft made a pass by Uranus, where it found ten new moons and discovered such oddities as a boiling ocean of water and winds as high as 450 miles (724 km) per hour.

Finally, in August 1989 *Voyager 2* flew just 2800 miles (4500 km) above far away Neptune, the planet's first flyby. Again, the flight revealed new moons (five), in addition to an atmosphere comprised of hydrogen and methane and winds of up to 680 miles (1100 km) per hour. Once the spacecraft had left Neptune and journeyed beyond the solar system, NASA redefined its rôle as the Voyager Interstellar Mission (VIM). In 2001 it had traveled 5,970,000,000 miles (9,600,000,000 km) since its launch twenty-four years earlier, carrying a gold-plated, 12 inch phonograph record encoded with music, sounds, spoken languages, and images from its home planet. *Voyager 1*, launched in September 1977 (two weeks after its sister

ship), actually reached Jupiter and Saturn before *Voyager 2* because of its more direct route. Because *Voyager 1* traveled close to Saturn's moon Titan, its trajectory led it straight from the vicinity of the ringed planet into the vastness of interstellar space, where it too joined the VIM mission.

By summer 2006, *Voyager 1* had reached a distance of more than 9,300,000,000 miles (15,000,000,000 km) from the Sun. Its instruments, as well as those of *Voyager 2*, continued to transmit data to Earth.

1. "Adm. Alan Shepard, Jr. Biography," *The Hall of Science and Exploration*, quoted in Tara Gray, "Alan B. Shepard, Jr.," p. 5; available from http://history.nasa.gov/40thmerc7/shepard.htm; Internet; accessed February 19, 2005.

2. Robert S. Kraemer, *Beyond the Moon: A Golden Age of Planetary Exploration, 1971–1978*, Washington, D.C. and London (Smithsonian Institution Press) 2000, p. 142.

5

The Shuttle
Spaceflight in a Time of Pragmatism

In 1969, an artist depicted post-Apollo hardware and operations as envisioned by President Nixon's Space Task Group. The system featured three main ingredients: a space tug, a shuttle orbiter, and a nuclear shuttle (or space station). To supply the space station, the orbiter would haul a cargo module from Earth, the tug would remove the cargo module and dock with it, and the loaded tug would fly to the nuclear shuttle.

A conception of the space shuttle, 1962. This proposal conceived of two parallel solid rocket boosters combined with three main engines on a shuttle orbiter. On a launch, at about 25 miles (40 km) altitude the boosters would fall to the ocean on parachutes, to be recovered and restored for future use. The shuttle, as envisioned, would later land on runways.

Less than two months after the safe return of the *Apollo 11* astronauts in July 1969—a time when Americans expressed their jubilation in the streets—President Nixon's Space Task Group on the post-Apollo era reported its findings to the nation. As a consequence of the triumphal lunar landing, the panel envisioned an expansive age of exploration, centered in part on the design and fabrication of large space stations.

But the president had other ideas. Nixon made it plain that his administration's plans for space exploration in the years following Apollo entailed practical measures, not heroic deeds. Indeed, the objective of lowering the cost of launching human beings and machines out of the atmosphere became the capstone of his program. Pressed to reduce government expenses due to the Vietnam War, Nixon left no doubt about the future. In replying to the Space Task Group report (entitled *The Post-Apollo Space Program: Directions for the Future*), the president wrote in March 1970:

We should work to reduce substantially the cost of space operations. Our present rocket technology will provide a reliable launch technology for some time. But as we build for the longer range future, we must devise less costly and less complicated ways of transporting payloads into space. Such a capability—designed so that it will be suitable for a wide range of scientific, defense, and commercial uses—can help us realize important economies in all aspects of our outer space program.[1]

The need for a new propulsion system occurred not just to a president intent on achieving fiscal prudence. Nixon's Space Task Group also recognized this reality. But its members posed the issue of access to space in a broader framework than that of the budget-minded chief

executive. They recommended as a first priority a space station that would orbit the Earth and serve as a base of operations to survey the nearby planets, especially Mars. To enable it, the group envisaged a vehicle that would maximize reusable parts and commonality in its design, and would serve as a low-cost supply train for the space station.

In fact, the concept of a reusable, inexpensive, multi-purpose launch system had originated many years earlier, the brainchild of neither the Oval Office nor the Space Task Group. Back in the early 1960s, as plans were materializing for the titanic *Saturn V* rocket, a small group of engineers in industry and government began to conceive of ways to lift heavy cargoes into space more economically than the classical methods adopted by Wernher von Braun and his associates at the Marshall Space Flight Center.

The individuals who advocated these new concepts grasped a fundamental political reality: members of Congress might approve lavish spending for a Cold War contest like Apollo, but eventually space exploration needed to stand (or fall) on its own merits. When that time came—presumably when the objectives of Apollo had been fulfilled—the infrastructure costs of future travel needed to be lowered radically if they were to avoid the axes of politicians prepared to either slash or eliminate further human spaceflight. An analogy posed by these technological insurgents illustrated their thinking. During the 1960s, an average American Airlines Boeing 727 accommodated 131 passengers and cost $4,200,000. If it flew only once (like the expendable *Saturn V*), each passenger needed to pay about $30,000. By the same logic, each astronaut aboard a *Saturn V* flight

EARTH ORBIT CARGO TRANSFER

MSFC-70-PD-4000-25B

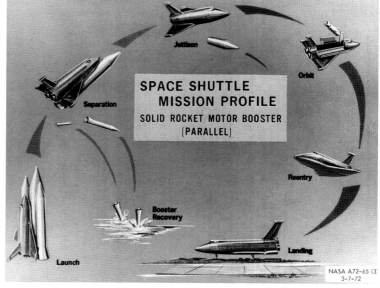

SPACE SHUTTLE MISSION PROFILE
SOLID ROCKET MOTOR BOOSTER
(PARALLEL)

NASA A72-65 (3)
3-7-72

↑ ↘ The research saga of the lifting bodies offered inspiration for the space shuttle concept. In contrast to the first generation of spaceflight, in which landings occurred at sea, NASA planners wanted to achieve piloted re-entry from space, in machines capable of landing on runways like standard aircraft. The wingless lifting bodies—flown at the Flight Research Center between 1963 and 1975— demonstrated one solution to this engineering hurdle. The M2-F3 (bottom picture), lineal successor to the humble, unpiloted M2-F1, demonstrated the capacity to fly supersonically, handle adequately, and land reliably like a conventional aircraft. So did the Air Force's X-24A (top picture, left) and the HL-10 (right), flanking the M2-F3 on Rogers Dry Lake, California.

needed to pay $60,000,000 to defray their ride into space.

The concept of a reusable launch vehicle that minimized single-use components received its first tests in 1961 and 1962. Engineers at Marshall immersed a Rocketdyne H-1 engine (the powerplant of the *Saturn IB*) in seawater, simulating a tumble into the Atlantic during recovery. They found that after a tear-down, inspection for corrosion, and reassembly, it performed perfectly well. But the narrower question of the overall design of a reusable system soon divided researchers into two camps. Some wanted a gigantic rocket with wings or a deployable chute that would fall to Earth after a launch to a "soft" splashdown, followed by refurbishment. These advocates favored projects like the Nexus rocket, a General Dynamics design of a single stage behemoth weighing 24,000 tons (24,384 tonnes) and capable of boosting 1000 (1016 tonnes) tons of payload into space. Others proposed an actual aircraft, taking encouragement from the contemporary X-15 rocket plane, which demonstrated the concept of at least limited reuse. This faction pursued advanced propulsion concepts, hoping to find a technology like the scramjet—a ramjet with supersonic internal airflows—that would enable single-stage-to-orbit flight. By 1963, however, it became clear that in the existing tight-fisted economy neither of these approaches held much promise.

But a compromise between the two schools offered some hope, and by the late 1960s one began to evolve. The convergence started at a NASA headquarters conference hosted by George Mueller, the agency's Director of Manned Space Flight, in January 1968. The attendees heard a novel idea proposed by Lockheed engineer Max Hunter. Eager to duplicate the efficiencies he knew from his days as a performance analyst at Douglas Aircraft, Hunter conceived the "stage and a half" concept: expendable and simple tankage for propellants,

connected to a core structure holding all the essentials for spaceflight. Once the fuel had been consumed shortly after lift-off, the tanks would fall away and descend into the ocean for later recovery. Meantime, the flyable main module envisaged by Hunter would undergo its mission and fly home for repair and eventual reuse. Representatives from General Dynamics presented an entirely different package, but one that was not incompatible with Lockheed's. They put their faith in a piloted aircraft like the X-15, of which the recently completed flight test program had proved relatively inexpensive. One advantage: its pilots would be able to take corrective actions not possible in automated systems. They recommended boosting this vehicle into space on the back of an Atlas missile.

Even though the proposals presented to Mueller by Lockheed, General Dynamics, and other contractors showed promise, they failed to provide a full answer to

←⋯↘ *Designed by engineers at Langley Research Center, the HL-10 lifting body (shown on Rogers Dry Lakebed with NASA pilot Bill Dana to the right and a B-52 mothership overhead) proved to have the best handling qualities of the early lifting bodies. Leaving an immense contrail, an HL-10 accelerates after being dropped from the B-52 bomber.*

the riddle of reusability. Still, even at this early date, Mueller began to refer to some hybrid designs as a space "shuttle", a space truck capable of transporting persons and goods to the planned orbiting space station on a scheduled basis.

Yet, this evolving innovation in spaceflight had sources of inspiration besides those of NASA headquarters, its contractors, and the X-15. For one, the lifting body program pursued at the NASA Flight Research Center (in close collaboration with NASA Ames and Langley) offered a flight-proven method of returning from space (see Chapter 3 for the origins of the lifting bodies). These bizarre-looking, wingless vehicles offered extraordinarily high lift-to-drag aerodynamic ratios, enabling pilots to land reliably, but at a steep angle of approach. Engineers at these three NASA centers collaborated on three basic lifting body prototypes and flight tested them between 1963 and 1972. First came the M2-F1, the brainchild of Flight Research Center engineer R. Dale Reed. Shaped like a gumdrop, the unpowered M2-F1 and its heavyweight successors M2-F2 and M2-F3 proved (after some hair-raising encounters with erratic lateral roll control) the potential of piloted return from space. The Langley team conceived the second type, the flat-bottomed HL-10 lifting body, flown between 1966 and 1970. Its pilots also experienced lateral roll control problems, but the remedial steps proved surer and more effective than the M2s. Then, in cooperation with the Air Force, from 1969 to 1971 research pilots flew the bulbous X-24A aircraft and found fewer flaws than with the earlier lifting bodies. Finally, appearing at the Flight Research Center between 1973 and 1975, the long, sharp-nosed X-24B (built literally over the frame of the X-24A) flew so well that pilots compared its handling qualities to those of the time-honored F-104 fighter. It was quite natural, then, that the contemporary lifting body programs should inform the concepts

of those engineers engaged in fashioning proposals for reusable spacecraft.

Another project, known as Air Force Dynamic Soaring (or Dyna-Soar), also exerted a significant influence. Like so much of modern aerospace, Dyna-Soar owed its existence to the shock of *Sputnik I*. A few days after the Soviet probe went into orbit on October 4, 1957, the USAF merged several initiatives to form Dyna-Soar. The idea originated in the fertile mind of German rocket scientist Eugen Sänger, and involved a winged capsule that would glide back to Earth in stages as it re-entered the atmosphere from space.

Dyna-Soar also intrigued the NACA, and later NASA. Their scientists and engineers joined others meeting at Ames less than two weeks after Sputnik to debate the many methods of launch and re-entry. Each had its supporters, but none won a consensus. As discussed in Chapter 3, the first involved a simple capsule (based on the blunt body concept of NACA Ames scientist H. Julian Allen), advocated by Langley's Maxime Faget. The second favored piloted lifting bodies. The third method—briefed by Langley's John Becker—envisioned a flat-bottomed hypersonic glide vehicle not unlike the X-15. Air Force officials folded Becker's blueprint into Dyna-Soar (the fourth option) and by the end of 1957 unveiled it as a small, single seat hypersonic boost-glide demonstrator, scheduled for initial test flights in July 1962.

Dyna-Soar progressed in three stages. It began by expanding the envelope of the X-15, flying the vehicle to speeds of 12,250 miles (19,700 km) per hour. During phase two, engineers proposed a two stage booster to propel Dyna-Soar to altitudes of 350,000 feet (107,000 m) and speeds of 15,000 miles (24,000 km) per hour before the long glide home. Its missions at this point focused on reconnaissance photography and attack on targets. Then, in a third incarnation, Dyna-Soar's planners foresaw a spaceplane capable of orbital flight, sophisticated reconnaissance, and even bombardment. John Becker and other Langley researchers contributed significantly to the project, both before and after a 1958 agreement with the USAF to provide technical advice to Dyna-Soar. The team of Boeing and Vought won the prime contract to design and fabricate the vehicle, now designated the X-20. But three years from a first flight and having already spent

$410,000,000 on the project, the Defense Department cancelled Dyna-Soar in December 1962.

Yet, the experiences, data, and techniques gleaned from Dyna-Soar and the lifting bodies figured prominently in the post-Apollo debates about NASA's direction. Maxime Faget played a decisive role among those engaged in the discussions. Above all, Max Faget liked simple and conservative designs. He garnered support for the Mercury capsule based on its relatively plain structure and operation, on its proven capacity to withstand the heat of re-entry, and on its ease of recovery. During the Apollo program, Faget initially opposed Dr. John Houbolt's Lunar-Orbit Rendezvous (LOR) method of landing on the Moon, presumably because of its inherent complexity. The hot-tempered Faget actually denounced Houbolt in a public meeting, exclaiming, "His figures lie. He doesn't know what he's talking about."[2] But, to his credit, Faget later reversed his position on LOR when studies proved that more direct approaches to the lunar surface offered little chance of success.

Into the late 1960s, Faget remained faithful to uncomplicated design, which he once again championed in his involvement with the space shuttle. When he entered the shuttle debate, the lifting body concept, then undergoing successful flight research at Edwards, seemed like the frontrunner among reusable candidates. But Faget all but took it out of play with his characteristically forceful arguments. He felt that the lifting bodies landed too fast (250 knots) for comfort or safety. Also, because the M2, HL-10, and X-24 fuselages themselves did the lifting, any aerodynamic problems that surfaced might require a redesign of the entire vehicle. Instead, Faget conceived of a two stage configuration comprised of an aircraft (the size of a 707 airliner) launched on the back of a huge, winged rocket (as big as a Jumbo 747). At the edge of space, the two vehicles would separate and the booster would return to Earth. Later, the spacecraft would maneuver for a landing similar to that of the X-15: nose high for much of the descent so that it would benefit from its blunt body shape. Then, at 40,000 feet (12,000 m), the re-entry speed would slow sufficiently for a nose-down approach, finally landing at a comfortable 130 knots.

But Faget's proposal led to vigorous opposition from engineers at the Air Force Flight Dynamics Laboratory.

---> *The needle-nosed X-24B, the last of the lifting bodies, proved to have the best handling of all. Flown at the Flight Research Center between 1973 and 1975, its rocket engine attained a speed of Mach 1.76 and an altitude of 74,100 feet (22,586 m). Collectively, the lifting body flight data (along with that of the X-15) gave the designers of the Space Shuttle strong assurance that astronauts could return home safely by runway landings.*

↑ *The YF-12 aircraft, Mach 3 plus cousin of the SR-71, featured delta wings, a planform found on other big, fast Air Force vehicles such as the XB-70 and B-58 bombers, as well as the planned Dyna-Soar. Delta wings also appealed to the designers of the Space Shuttle Orbiter.*

They objected to its straight-wing planform on the basis of the likelihood of unpredictable swings in the center of lift as the vehicle descended through different speeds on its return flight. The answer—chosen for Dyna-Soar and actually employed by the USAF on such fast, big aircraft as the XB-70, the B-58 bomber, and the SR-71 reconnaissance plane—lay in delta wings. This configuration enabled greater lift at hypersonic speeds and counteracted the migrations in the center of lift by incorporating elevons in the design.

As the technical wrangling continued through 1969 and 1970, the very reason for building a shuttle became a victim of the budget ax. Until this time, the Nixon administration had conceived of the shuttle as a supply truck for the planned space station, a heavy lift system

capable of hauling big structural pieces of the station in a cavernous cargo bay. When the administration abandoned the space station during 1970, the shuttle lost its reason for being, leaving NASA officials in desperate need of a patron to share its capabilities and cost. They found a backer in the Department of Defense, but salvation came at a price. Fearful of diminished capacity for military launches during the 1980s, U.S. Air Force negotiators entered talks with the space agency. NASA representatives had little bargaining power. They needed the Defense Department's military and intelligence business in order to persuade Congress of the shuttle's continued validity. On the other hand, the Air Force— restricted to lifting objects no wider than 10 feet (3 m) on its existing Titan rockets—required a cargo bay of

60 by 15 feet (18.3 by 4.6 m) for larger future spacecraft. The service made additional demands. It insisted on a vehicle capable of lifting up to 40,000 pounds (18,000 kg) into low polar orbit, and held out for a delta wing design and a crossrange of 1100 nautical miles (2037 km). To keep the project alive, NASA conceded all these points at a meeting with USAF officials in Williamsburg, Virginia, in January 1971.

Winning Department of Defense collaboration played a decisive role in NASA's achieving congressional approval of its reusable launch vehicle, based upon the demands of national defense. But the space agency also hoped for presidential support. President Nixon often expressed admiration for the astronauts, felt the nation needed the example of their bravery, and realized that success in space translated into diplomatic leverage for the United States. Unlike President Kennedy, however, he sought fiscal prudence at the same time as heroism. Moreover, while the martyred president couched space travel in the poetic language of discovery, Nixon never allowed his enthusiasm to overcome his practicality.

It is not surprising, then, that Nixon's endorsement of the Space Shuttle should originate with the Office of Management and Budget. Here, in a federal agency almost feared for its fiscal stringency, the shuttle found an unlikely champion. During the summer of 1971 NASA Administrator James Fletcher made the shuttle case to Deputy Director of Management and Budget (and future Secretary of Defense) Caspar Weinberger. Having failed to win other administration converts, Fletcher at last found a friend.

Weinberger wrote a memo to the president, arguing that at the apex of the American lunar triumph, the painful reductions in NASA's budget needed to be reconsidered. He presented a case for NASA as an investment in the future and for undertaking activities that would have inspirational value. He asked the president to consider not just social issues or the cost of fighting the Vietnam War. The nation, Weinberger said, "should be able to afford something besides increased welfare, programs to repair our cities, or Appalachian relief . . .." NASA and the proposed shuttle represented one such future-looking endeavor, "offer[ing] the opportunity . . . to secure substantial scientific fall-out for the civilian economy [and] at the same time [enabling] large numbers of valuable (and hard-to-employ elsewhere) scientists and engineers [to be] kept at work on projects that increase our knowledge of space. . . ." Weinberger warned that a full retreat from the space program implicit in the impending budget threatened all the nation had gained—strategically, scientifically, and psychologically—from Mercury, Gemini, and Apollo:

Recent Apollo flights have been very successful from all points of view. Most important is the fact that they give the American people a much needed lift in spirit (and the people of the world an equally needed look at American superiority). Announcement now, or very shortly, that we are canceling Apollo 16 and 17 (an announcement that we would have to make very soon if any real savings are to be realized) would have a very bad effect, coming so soon after Apollo 15's triumph. It would be confirming in some respects, a belief that I fear is gaining credence at home and abroad: That our best years are behind us, that we are turning inward, reducing our defense commitments, and voluntarily starting to give up our super power status, and our desire to maintain world superiority.[3]

Nixon's simple note on Weinberger's memo, "I agree with Cap," constituted perhaps the most off-handed birth certificate given by any president to a public project of such magnitude. Always the realist, however, Nixon had another motive besides the high-minded ones advanced by Caspar Weinberger. About to enter a re-election campaign, the president realized the political benefit of being able to promise thousands of new, high paying jobs in key states as a result of shuttle work.

Be that as it may, James Fletcher received an immediate go-ahead in January 1972 after a visit to the president at his home in San Clemente, California. During the initial phase, NASA needed to select a prime contractor for the shuttle, now known in all its complexity

←⋯ ↑ *While NASA pursued the Mercury and Gemini projects, the U.S. Air Force undertook a hybrid, combining an aircraft not unlike the X-15 with a Titan II booster. The result—known as Dynamic Soaring, or Dyna-Soar—began in 1957, and had it been completed, might have looked like the artist's conception shown here. Langley's John V. Becker, one of the fathers of the X-15, contributed significantly to Dyna-Soar research. Like the lifting bodies, Dyna-Soar also influenced the design of the Space Shuttle.*

Maxime A. Faget

Maxime A. Faget was born in 1921 in Stann Creek, British Honduras. He attended San Francisco Junior College and in 1923 received a bachelor's degree in mechanical engineering from Louisiana State University. Faget served during World War II as a naval officer, assigned to the Submarine Service. He left the military in 1946, joined the NACA, and for the next twelve years worked at the newly established Pilotless Aircraft Research Division (PARD) in Wallops Island, Virginia. Here, he and others collected aerodynamic data from the rocket-propelled aircraft and missiles flying at hypersonic speeds in the skies over Wallops. Faget's talents became recognized at PARD, and from 1950 to 1958 he became the head of the performance and aerodynamics branch. His assignments at PARD prepared him well for the American leap into space after the launch of *Sputnik I*. During discussions of spacecraft types that occurred after Sputnik but before the formation of NASA, he became a leading proponent of an uncomplicated, non-lifting, blunt body vehicle that would re-enter the atmosphere on a ballistic path. His concept formed the basis of the Mercury space capsule, including the use of the parachute for final leg of the flight. Others adapted his ideas once NASA had supplanted the NACA. One of the original thirty-five members of the Space Task Group headed by Robert Gilruth at Langley, Faget was chief of Langley's Flight Systems Division between 1958 and 1961. Then, with the creation of the Manned Space Center (later the Johnson Space Center) Faget moved to Houston and took on the role of Director of Engineering and Development, a position he held for twenty years. During this period he oversaw not only much of the design and analysis involved in the Apollo spacecraft but also the construction of many of Johnson's laboratories and, finally, the development of the Space Shuttle. Faget retired from NASA in 1981, founding (and serving as CEO of) Space Industries, Inc., a company that manufactures equipment supporting a variety of shuttle experiments. Faget died in 2004.

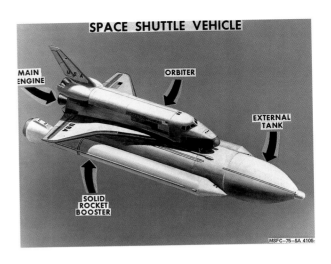

SPACE SHUTTLE VEHICLE

MAIN ENGINE — ORBITER — EXTERNAL TANK — SOLID ROCKET BOOSTER

MSFC—75—SA 4105-

as the Space Transportation System (STS). A Request for Proposals left NASA in March, and by May four firms had replied: North American Rockwell, McDonnell Douglas, Lockheed, and Grumman. A source selection board consisting of space agency and USAF officials reviewed the four proposals over the weeks that followed, and ranked the results.

Then, after preliminary assessments of the competing manufacturers, three NASA employees—Administrator Fletcher, Deputy Administrator George Low, and Richard McCurdy—convened on the morning of July 26 to choose the winner. Lockheed ranked last. Only a little more expensive than North American's proposal, the Lockheed system weighed more than the others, seemed unduly complex, required a fast landing speed, and appeared to be based on unrealistically low costs. More telling, the Lockheed engineers had never built a piloted spacecraft. McDonnell Douglas placed third. Although McDonnell had fabricated both the Mercury and Gemini capsules, and Douglas Aircraft had built Skylab, the merged company failed to make a persuasive and integrated case, giving the impression of two separate proposals from two separate firms. The high cost and technical weaknesses also bothered Fletcher, Low, and McCurdy. Grumman made a good showing, based on its partnership with North American in building the Apollo spacecraft. Its fine design and mastery of the details impressed the three

←---- *Pilotless Aircraft Research Division engineer Maxime Faget collaborated with John V. Becker on the fundamental X-15 design.*

↑ *An illustration of the Space Shuttle in its final form, 1975. Marshall Space Flight Center led a team of aerospace industries in the development, test, and fabrication of such pivotal shuttle components as the main engines, the external tank, and the solid rocket motors and boosters.*

----→ *NASA Ames also made major contributions to the shuttle, illustrated in this photograph of a one-third scale shuttle orbiter being tested in the Ames 40 by 80 foot (12 by 24 m) full scale wind tunnel in 1975.*

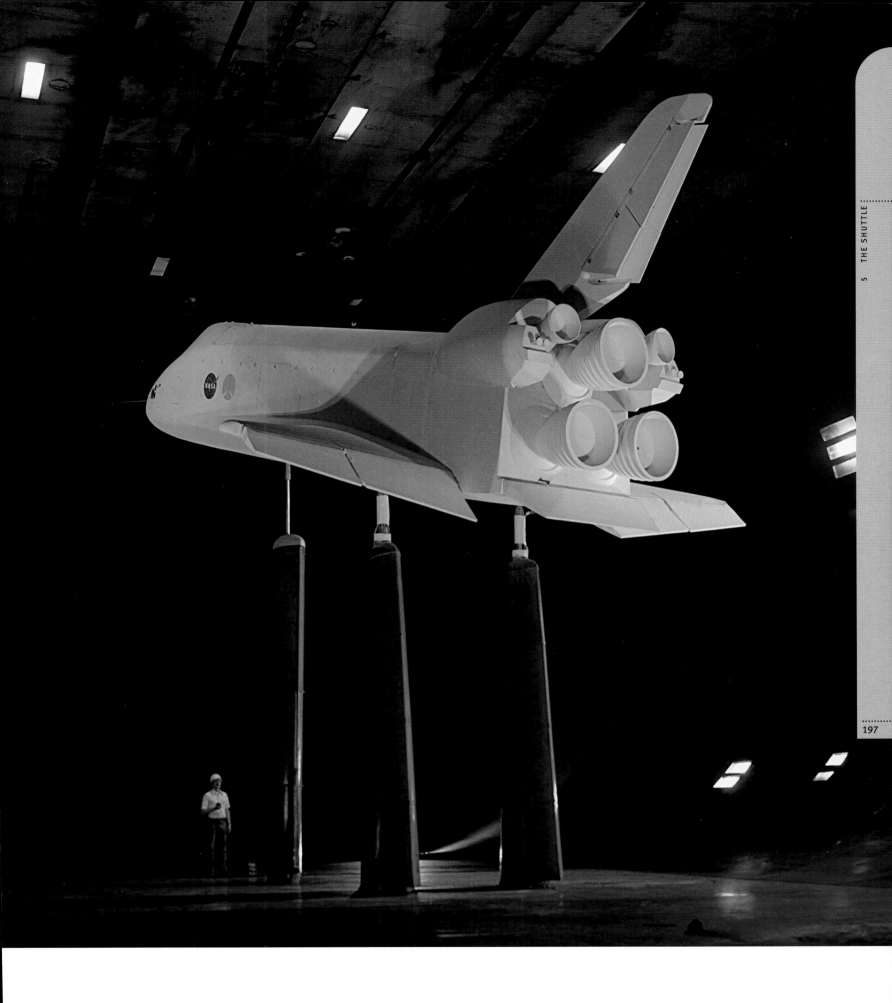

Amidst a throng of onlookers, the first of the colossal Shuttle external tanks rolls out of NASA's Michaud, Louisiana, Assembly Facility in September 1977. From there, it would undergo tests at Stennis Space Center in southern Mississippi.

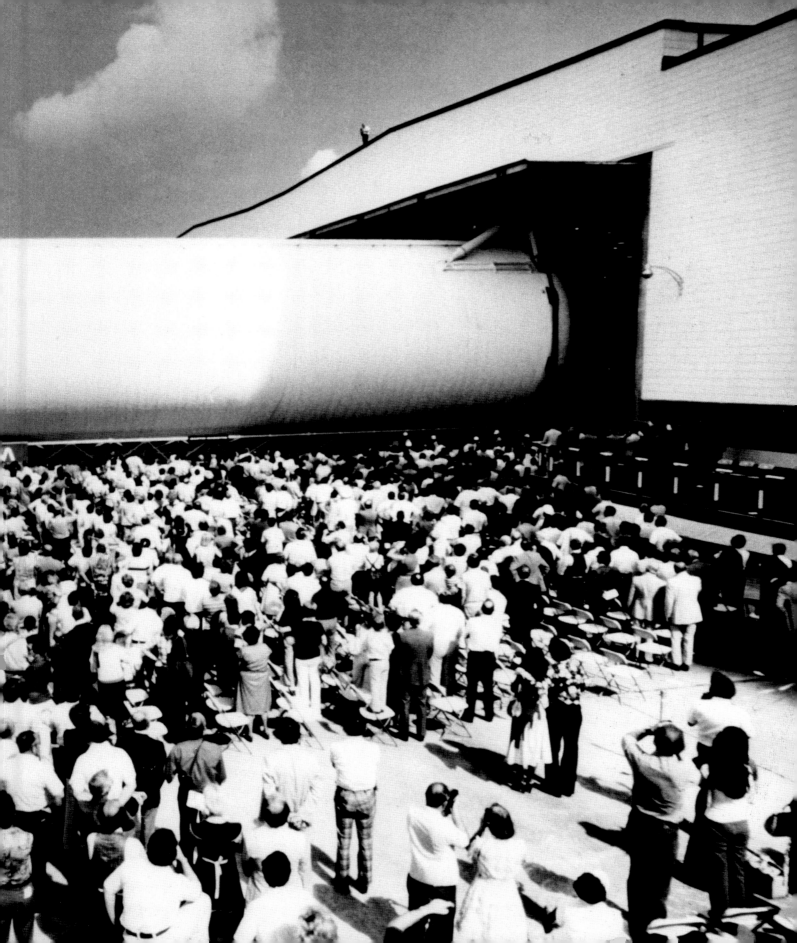

[Left] *Rockwell International's Rocketdyne Division fabricated the Shuttle main engines under contract to Marshall. In this image a machinist drills holes into the main engine's injector body, the corridor through which propellants flowed towards the combustion chamber.*

[Below left] *One of the essential components of the Space Shuttle orbiter, the Remote Manipulator System (RMS)—or Shuttle arm—served as a hoist, a crane, and an astronaut platform during repair and servicing missions. Its versatile end effector fitting is visible in this picture.*

[Below] *By March 1978, the (by now) familiar orbiter profile began to be seen outside NASA's assembly buildings. At Marshall, it rolled across the center on its way to the first vertical mating with the external tank and the solid rocket boosters.*

[Opposite] *In order to determine the actual aerodynamics of the shuttle orbiter in flight, program managers at Johnson Space Center asked their counterparts at the Dryden Flight Research Center (formerly the Flight Research Center) to conduct a series of experiments. During 1977 five Approach and Landing Tests occurred at Dryden. This picture shows Fred W. Haise, commander (left in picture), and C. Gordon Fullerton in the cockpit of the Shuttle* Enterprise *before its mothership—a Boeing 747 carrier aircraft—released it for an unpowered flight to the runway at Edwards Air Force Base. Also, a picture of research pilot and astronaut Gordon Fullerton.*

↑ ⟶ *Three initial phases of the Approach and Landing Tests. First,* Enterprise *is released from the back of the 747 (top). Second, the orbiter separates from the carrier aircraft. Third, the orbiter glides back to Rogers Dry Lake.*

men, but the management aspects failed to satisfy them. Moreover, the price seemed high and the firm admitted that it needed to swell its ranks hastily in order to meet the schedule.

Among the four, North American Rockwell made the most convincing presentation, although it too had significant flaws. Fletcher and the others suspected that its crew cabin might pose construction difficulties. But the North American team submitted a credible low bid, promised the lightest vehicle, and presented a highly maintainable system. By the afternoon, Fletcher, Low, and McCurdy had chosen North American to fabricate the orbiters and to oversee the integration of their overall STS conception.

In time, Rockwell's winning drawings became the stuff of history, transformed into iconic images familiar to television and movie screens, newspapers, and magazines the world over. The Rockwell STS concept featured a

gigantic stack of four parts. It consisted of a delta winged aircraft (the orbiter) 126 feet (38 m) long, 56 feet (17 m) tall at the tail, 84 feet (26 m) across its wingspan, and weighing about 150,000 pounds (68,000 kg); twin reusable solid rocket boosters 185 feet (56 m) in length each and weighing 1,293,000 pounds (587,000 kg) apiece; and, finally, a massive, expendable liquid fuel tank 206 feet (63 m) long and weighing over 1,638,000 pounds (743,000 kg).

Despite their elation, the engineers and scientists at North American also realized the immensity of their task. In order to make real the image of the massive Shuttle rising from the launch pad, countless formidable technical hurdles needed to be surmounted, in part because the new Space Transportation System constituted such a clean break from past space practices and such a daring technological advance. The small spacecraft of the early U.S. program landed in the ocean on a parachute and required recovery by the Navy. The Shuttle orbiter—the size of an airliner—was to glide home from space at hypersonic speed and land like an aircraft on a hard runway. The capsules of the early period carried no more than three inhabitants in cramped conditions. The Shuttle orbiter would accommodate as many as ten in relative comfort and have room for a bus in its cargo bay. But the most important distinction of all related to the Shuttle's partial reusability, the feature that had prompted its conception in the first place and represented an attempt to lower drastically the cost of spaceflight in a single stroke.

In winning the $2,600,000,000 competition, the prime contractor agreed to design and fabricate two spaceworthy orbiters, a full-scale test spacecraft, and a main propulsion test article; integrate the orbiters with the other parts of the Space Transportation System; and support shuttle operations at Kennedy Space Center during the first two years of operation. As North American's team attempted to satisfy these terms, in August 1973 a NASA source selection board picked Martin Marietta Corporation to design and manufacture the external tank. Thiokol Propulsion of Utah won the award to develop the twin solid rocket boosters.

As the project progressed, the contractor found it necessary to make subtle but noteworthy modifications in the orbiter's aerodynamic design, as well as undertake significant weight reductions, based in part on some 46,000 hours of testing in NASA, Air Force, and university wind tunnels. These experiments resulted in a somewhat more compact spaceplane. To reduce the vehicle's empty weight, NASA engineers made a trade-off: they accepted a faster rate of speed during landings, in exchange for smaller, lighter wings (shaped in a double-delta configuration, swept 79 degrees in the forward position and 45 degrees for the rest). They also asked their North American counterparts to build a more blended wing–body planform than that originally conceived in order to diminish re-entry heating on the sides of the fuselage. In addition, in light of the large, flat areas on the orbiter's surface, the project team decided that special care needed to be taken to fabricate the heat-resistant tiles (which shielded the orbiter from the intense heat of re-entry) as uniformly as possible. These and other important considerations resulted in an orbiter that was 3 feet (90 cm) shorter in length than previously, and endowed with a new airfoil design, a smaller nose radius, and a more flowing nose–body section. Later adaptations involved the installation of recessed thermal

The approach and landing part of the ALT tests. The orbiter, with a chase plane in pursuit, turns to line up its approach; Enterprise suspended above the runway; and the spacecraft rolling out after landing.

glass at the hatches, windscreens, and observation windows.

Less apparent changes also occurred. The space agency decided to depart from the initial concept of reaction control machinery cloaked behind opening and closing doors, and instead requested an exposed system that would be better able to withstand the rigors of ascent and descent, thus reducing both cost and complexity. Project officials also decided to dispense with the parachute braking system incorporated in the test orbiter, feeling that the immense lakebed at Edwards Air Force Base—where the flight trials occurred—offered ample room for long rollouts. Finally, engineers altered the lengths of both the external tank and the solid rocket boosters slightly. These and many other reassessments of North American's initial concept underwent a Critical Design Review in February 1975, at which the contractor won approval to proceed.

In order to test the airworthiness of the orbiter design at the earliest possible opportunity, North American pressed ahead with the fabrication of the non-operational vehicle as it worked on the two destined to go into space. Christened *Enterprise* in honor of the *Star Trek* ship of the same name, the testbed was wheeled out of

the manufacturer's plant in September 1976. Meanwhile, at the NASA Dryden Flight Research Center the staff prepared for the ship's arrival. Transported at a stately pace on a flatbed truck, *Enterprise* made the day-long, 36 mile (56 km) passage on January 31, 1977. Onlookers snapped photos as the big spaceplane left the cavernous Rockwell facility in Palmdale, inched northward over barren desert to Lancaster, and then entered Edwards Air Force Base.

NASA officials had been preparing for this moment for years. Astronaut Deke Slayton at Johnson Space Center, the project director of the upcoming trials at Dryden, took command of a series of flights known as the Approach and Landing tests. In these flights, the orbiter—mounted on a large aircraft and released high above the desert—made a number of piloted, unpowered glides to Rogers Dry Lake below. During the drops of the vehicle (originally eight, but reduced by Slayton to five in order to expedite development), NASA engineers would plot the flight maneuvers, take note of pilot impressions of handling, and consult the instrumentation data in order to determine the spacecraft's aerodynamic mettle.

To lay the groundwork for these pivotal flights, NASA

pilots compared the flying qualities of two candidate carrier aircraft: the 747 Jumbo and the C-5 Galaxy military cargo plane. The airliner seemed to offer cleaner separation between orbiter and mothership, and NASA procured an older, obsolete model for the experiments. Meanwhile, at the far end of the Dryden flightline, a strange tower rose during 1976, consisting of two 100 foot (30 m) vertical structures linked 80 feet (24 m) up by a horizontal arm. Known as the Shuttle Mate–Demate Device (MDD), this towering piece of steel scaffolding enabled the orbiter and the Shuttle Carrier Aircraft (as the 747 became known) to be attached for the upcoming ferry flights to Kennedy Space Center.

After eight ground and airborne tests of the 747/orbiter stack between February and July 1977, the first actual Approach and Landing flights occurred in August. On the initial one, Fred Haise of *Apollo 13* took control of the spacecraft following the release of the orbiter at 22,000 feet (6700 m), and although it maneuvered satisfactorily and landed safely, the second

of four redundant computers shut down and the vehicle landed a mile (1.6 km) beyond the designated landing point. Two additional flights occurred in September (the second also commanded by Haise), and one in October. Again they missed the landing targets, and encountered brake chattering and significant braking ineffectiveness.

The last mission took place on October 26, when Commander Haise flew *Enterprise* once more. But on this occasion, all did not go well. Perhaps feeling the pressure of one last chance to make good, Haise tried hard to land close to the targeted 5000 foot (1524 m) marker. Complicating his challenge, *Enterprise* flew for the first time with its tailcone removed, causing a big difference in its aerodynamics. Moreover, Prince Charles joined the many observers that day, adding to media interest and perhaps to Haise's jitters.

At any rate, as Haise approached for the landing, he worked the controls in a tight, intense manner, with the result that he came in nose high and a little fast. He applied the brakes. Then he used the control stick to alter

↑ ⋯→ Research on the Navy F-8 Crusader (in flight over Edwards Air Force Base) enabled the solution to Shuttle Orbiter instability. Digital fly-by-wire occurred first on the F-8, but not before it had been tested and refined on an "Iron Bird" simulator with layout and features much like the real vehicle. Engineers adapted the software developed for the F-8 to the orbiter spacecraft.

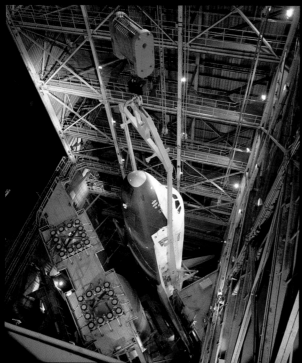

[Left] *Between the end of the Approach and Landing Tests in October 1977 and the first flight of the Shuttle* Columbia *in April 1981, the complicated machinery that comprised the entire Shuttle structure—the Space Transportation System (STS)— underwent untold numbers of tests. In this instance, the orbiter is lifted on a hoist from Marshalls's Dynamic Test Stand after undergoing a Mated Vertical Ground Vibration Test.*

[Bottom left] *Another view of the ground vibration test showing the orbiter* Enterprise *being lowered into the Marshall test stand. Much of the research on the Shuttle had to be simulated in wind tunnels and on ground equipment such as this due to the immensity of the STS and the inherent costs.*

[Right] *Another aspect of the STS that needed to be perfected, was the Mate–Demate Device, used to mount the orbiter onto, and to dismount it from, the 747 carrier. Kennedy Space Center and Dryden Flight Research Center both had Mate–Demate Devices for the anticipated times when weather or other circumstances would prompt Shuttle landings in California and subsequent ferry flights to Kennedy.*

↓ *A researcher at the Marshall Space Flight Center holds a small orbiter model in the Trisonic (subsonic, transonic, and supersonic) tunnel. Used to test the integrity of rockets in re-entry and launch conditions, it has assessed vehicles from the era of the Jupiter-C to the present.*

⋯⋯↘ *One of the most impressive sights in the history of rocketry, the massive Shuttle stack—the orbiter* Enterprise, *the external tank, and the solid rocket boosters—is shown being rolled to the launch pad on the Shuttle transporter.*

Enterprise's sink rate, causing swings in the vehicle's elevons, and in turn, prompting pitch up (nose to tail oscillations). Haise made inputs to the onboard computer in an attempt to damp the motions. Just as *Enterprise*'s wheels touched down, the aircraft bounded upwards as the flight control surfaces responded belatedly to Haise's commands. Then, four seconds of hair-raising pilot induced oscillations (or PIOs, violent rolling motions from wing to wing) shook the vehicle. An instant before contacting the ground for the second time, pilot Gordon Fullerton in the second seat yelled, "Hey, let loose." Haise took his hands off of the controls, and the vehicle recovered for the touchdown, averting a likely crash.

The orbiter's instability on approach had been a crucial discovery. Researchers at Dryden, fresh from a series of successful digital fly-by-wire flights aboard a Navy F-8 Crusader, collaborated with Johnson—the lead center for shuttle development—to devise software that "filtered"

this problem in the spacecraft's flight characteristics. But this fix happened only after extensive simulated shuttle-style landings aboard such varied aircraft as a Navy F-8 and the Mach 3 YF-12 research aircraft. These flights suggested that Haise's wild ride resulted in part from a delay in the mechanical response to the computer commands during the final moments of the fifth flight. The software filtering techniques that suppressed (but did not eliminate) the PIO and the pitch-up tendencies experienced by Haise continued to be investigated during the late 1970s, and Houston eventually incorporated them into the orbiters in 1979.

During the same year, the first of the operational orbiters left the North American hangar in Palmdale. In March 1979, *Columbia* (named for a venerable U.S. Navy frigate launched in 1836, which circumnavigated the world) flew aboard the 747 carrier aircraft to Kennedy Space Center in preparation for the Space Shuttle's maiden launch. As it underwent preparations, Congress approved the fabrication of three other orbiters, all, like *Columbia*, named for seagoing vessels of the past. *Challenger* commemorated a U.S. Navy vessel that undertook voyages of exploration in the Pacific and Atlantic during the 1870s. *Discovery* recalled a ship captained by Henry Hudson during the hunt for the Northwest Passage (1610–11), and the fabled journey of James Cook during the 1770s to the Hawaiian Islands and western Canada. Finally, *Atlantis,* chosen for the fourth orbiter, had more recent and humble origins, derived from a venerable two-masted ketch that sailed the seas for the Woods Hole Oceanographic Institute from the 1930s to the 1960s.

But even as these vehicles rose on the assembly floor at North American Rockwell, some wondered whether the economical, reusable space truck envisioned by President Nixon and anticipated by its early designers might be realized. NASA projected outlays of about $500,000,000 to $600,000,000 for each of the new orbiters; in reality, they cost over $1,000,000,000 apiece. Still, these spacecraft and their supporting engines and fuel tanks represented a monumental technological accomplishment. Indeed, the expendable Apollo–*Saturn V* combination that had represented state-of-the art American rocketry only a few years earlier now seemed

passé compared to the big space plane and its two immense, reusable solid rocket boosters. When the entire massive STS stack stood vertically on Launch Pad A at the Kennedy Space Center in April 1981, it embodied one of the greatest feats of human design and construction. Not surprisingly, it engendered immense public curiosity.

All seemed ready for launch on the tenth of that month until technicians detected a software problem. Then, two days later, *Columbia*—weighing nearly 4.75 million pounds (2.2 million kg) with payload—shook and roared as its three main engines fired. As the solid rocket boosters came to life, each contributing 2.65 million pounds (1.2 million kg) of thrust, the gigantic structure lifted off, staying intact until about two minutes after lift-off. At that moment, the boosters detached from the orbiter and fell away on parachutes, to be retrieved from the Atlantic Ocean by ships and refurbished for reuse. The orbiter's main powerplants burned for about eight minutes, fed by fuel from the external tank, which, when empty, separated and plunged to Earth, disintegrating as it re-entered the atmosphere. Finally, *Columbia*'s orbital maneuvering engines fired

⟨···· ····⟩ *The first Shuttle flight— STS-1, flown by the orbiter Columbia—lifts off of Kennedy Pad 39A on the morning of April 12, 1981. After orbiting 36 times, astronauts John Young and Robert Crippen landed 54 hours later on Runway 23 at the Dryden Flight Research Center, Edwards Air Force Base, California.*

⟍ *A view of the astronauts leaving* Columbia *after STS-1 completed its mission. Robert Crippen walks out of the orbiter as Dr. Craig Fischer, Chief of Medical Operations at the Johnson Space Center, follows. John Young and George Abbey, Director of Flight Operations at Johnson, stand at the foot of the stairs.*

[Left] *James Beggs became NASA's sixth Administrator (sworn in by Vice President George Bush) on July 10, 1981. He began his term shortly after the Shuttle started flying and left in December 1985, just before the* Challenger *disaster. Beggs had been executive vice president and director of General Dynamics Corporation. James Fletcher returned to NASA after Beggs's departure.*

[Above] *One of the first female astronaut class, Dr. Kathryn Sullivan, appears here in a high altitude pressure suit in front of a NASA WB-57F reconnaissance aircraft. Flight in this vehicle gave her training for the Shuttle, on which she flew in STS-41G, STS-31, and STS-45 as a mission specialist.*

twice: once to put the spacecraft into orbit, the second to stabilize its trajectory. In all, its five engines (three on the orbiter, one each on the solid rocket boosters) developed 7 million pounds (3.2 million kg) of thrust, boosting this airliner-sized spacecraft into orbit in a mere ten minutes.

Essentially, Commander John Young (veteran of *Gemini 3*, *Apollo 10*, and *Apollo 16*) and pilot Robert Crippen (formerly associated with the USAF's Manned Orbiting Laboratory) took *Columbia* on a two-day shake down cruise. Its main payload consisted of developmental flight instrumentation equipment to record the temperatures, pressures, and accelerations at different points on the spacecraft. After checking the orbiter's computers, bay doors, and various systems, the astronauts declared everything to be satisfactory. On April 14 they employed the maneuvering rockets once more to reduce the spacecraft's speed below the 17,500 miles (28,200 km) per hour required for orbit. Young and Crippen touched down at NASA Dryden Flight Research Center on Runway 23, co-located on Edwards Air Force Base, California, the planned landing site of all of the initial Shuttle flights. Technicians inspecting the orbiter did notice some disturbing wear and tear after the thirty-six trips around the planet and the intense re-entry: sixteen heat-resistant tiles lost, 148 damaged, and some minor ill-effects from overpressure caused by the boosters at ignition. While worrisome, these results did not seem very consequential at the time, although the second flight of *Columbia* (November 1981) returned with twelve damaged tiles, and

Sally K. Ride

Born in Los Angeles, California, in May 1951, Sally K. Ride, the first American woman to fly in space, showed early affinities for athletics and science. A fine tennis player, ranked eighteenth on the junior circuit, she also possessed an aptitude for the intellectual life, inherited perhaps from her father Dale, a professor at Santa Monica Community College. She earned a B.S. in physics, a B.A. in English literature, and finally a doctorate in physics, all from Stanford University. Upon graduation in 1978, she responded to calls by NASA for mission specialists, the first class to enroll women in the astronaut corps. Dr. Ride went aloft on June 18, 1983, aboard *Challenger* on the seventh Space Shuttle flight. The largest shuttle crew to date (five), they launched two commercial satellites (*Anik C-2* and *Palapa B-1*) and also practiced using the remote manipulator arm, releasing and capturing *SPAS-01* (a shuttle pallet satellite that carried ten experiments). Dr. Ride later served on the [William] Rogers Commission that investigated the *Challenger* accident, and in 1986 and 1987 she chaired a NASA task force to restructure the agency and return it to spaceflight. She left NASA in 1987 to join the Center for International Security and Arms Control at Stanford University, but returned to scientific pursuits two years later when she became the director of the California Space Institute and professor of physics at the University of California at San Diego.

←···· *The first American woman in space, Dr. Sally Ride, served as a mission specialist on STS-7. Here she monitors the control panels on the pilot's deck as a flight procedures notebook floats in front of her.*

←···· ←···· *Because the Space Shuttle flew relatively often and carried far more human beings than the earlier spacecraft, women and minorities faced better odds of joining the astronaut corps. The rise of the women's rights, as well as the civil rights movements, also opened doors. Pictured in 1978, the first female astronaut candidates undergoing water survival school at Homestead Air Force Base, Florida: (left to right) Sally Ride, Judith Resnik, Anna Fisher, Kathryn Sullivan, and Rhea Siddon.*

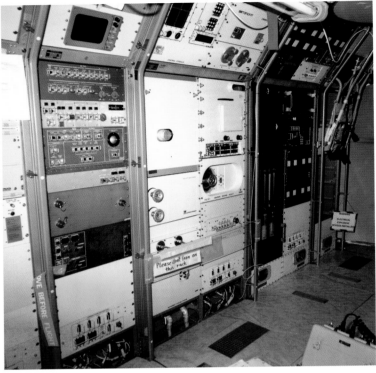

the third (March 1982) resulted in thirty-six lost and nineteen damaged.

After the success of the first shuttle mission, the press and NASA itself dared to believe that an era of routine access to space had dawned. Subsequent shuttle achievements seemed to bear out the optimism. Incredibly, the STS, which represented not one but a cluster of new technologies, flew its first five years without a serious incident, its crews suffering no injuries and no fatalities. It also proved its worth in the pursuit of knowledge. On the ninth Shuttle flight (November 1983) the STS's scientific value became evident with the launch of the European Space Agency's *Spacelab I*, a small-scale laboratory carrying seventy-two experiments dealing with physics, astronomy, materials, life sciences, and Earth sensing. Its six crewmembers (the largest number to date) required consecutive twelve hour shifts for the duration of the mission to finish the lab's work. *Spacelab* made three more flights aboard the Shuttle before 1986.

Moreover, the Space Shuttle enabled many to experience space travel. By 1985, seven persons per flight had become commonplace and during the first twenty-four missions (up to January 12, 1986) the STS had lifted 125 individuals into the heavens. Truly, the carrying capacity of the Shuttle orbiter democratized spaceflight, creating opportunities for women and minorities hitherto unknown. For example, Dr. Sally K. Ride, a physicist from California, flew aboard STS-7 in June 1983, surmounting the gender barrier that had been present in the U.S. space program since its inception.

But in spite of the initial programmatic, and later operational optimism, after five years it had become clear that the orbiters failed in the role of economical space transports that had been hoped for in the beginning. They required months—not days or weeks—of inspection and repair between flights. They required a vast maintenance and engineering staff to keep them aloft. Clearly, by the mid-1980s, the STS had earned its reputation as a reliable form of heavy launch and as a technology that inspired awe and appreciation. But its reason for being (frequent and cost-effective spaceflight) had already proven illusory. In fact, an unrealistic desire to press the Shuttle to meet these expectations resulted in one of the worst—certainly one of the most horrifying—technological catastrophes in American history. Until that point, the Shuttle had flown twenty-four times and nothing in the earlier experiences prepared the space agency for the events of January 28, 1986.

ON THE NINTH SHUTTLE FLIGHT (NOVEMBER 1983)
THE STS'S SCIENTIFIC VALUE BECAME EVIDENT WITH THE
LAUNCH OF THE EUROPEAN SPACE AGENCY'S *SPACELAB 1*.

Spacelab had a long genesis. Conceived in 1973 as a European Space Agency contribution to Shuttle flight, a succession of Spacelabs operated from the cargo bays of the orbiters for week-long scientific investigations. Spacelab 1, pictured in the cargo bay of Columbia, *went into orbit on November 18, 1983.*

[Left] *After the success of STS-1, the flight of STS-2 awakened presidential interest in the space program. At the Mission Operations Control Room at Johnson, President Ronald Reagan (top picture) looks up as Christopher Kraft calls attention to the projection plotter showing the track of* Columbia. *Vice President George Bush also came to Johnson for STS-2, speaking after breakfast with Christopher Kraft.*

[Below and opposite] *After the launch of STS-2, Flight Director Neil Hutchinson watches the data on his cathode ray tube monitor at the Johnson Mission Operations Control Room. After a two day mission, on November 16, 1981,* Columbia, *piloted by Richard Truly, makes its final approach to the Edwards Air Force Base runway (far right). At the end of STS-2, Truly (near the bottom of the stairs) and Joseph Engle descend the steps to the runway at Dryden Flight Research Center. Nine days later,* Columbia *arrives at Kennedy Space Center and technicians separate the orbiter from the 747 carrier (top photograph).*

The Space Shuttle needed far more maintenance and repair than its designers had hoped. Although not rated for spaceflight, the Shuttle Enterprise did serve a useful purpose for engineering fit-checks after the Approach and Landing Tests in 1977. Here, it undergoes weight and balance evaluations at Dryden Flight Research Center. The Smithsonian Institution's National Air and Space Museum accepted Enterprise into its collection and it went on display (in restored condition) at the Udvar-Hazy Annex in 2004.

All appeared well as Challenger lifted off on January 28, 1986. But on closer inspection (note a small puff of grey-brown smoke emanating from the solid rocket booster, at a point opposite the orbiter at the "U" in United States) dire events could be seen unfolding. Due to uncommonly cold overnight temperatures, one of the rubber O-ring seals at a joint in the solid rocket booster had become brittle, causing a leakage of gases which cut open the booster, resulting in this initial puff of smoke. But as smoke became flame, the flame acted as a blowtorch on the surface of the external tank.

Planned for launch on January 22, 1986, STS-33—flown aboard *Challenger*—had been postponed three times as a result of inclement weather, competing demands on the KSC workforce, and last-minute changes in the final integrated simulation. Then, on January 27, the new date of launch, a faulty handle on the external crew hatch slowed down preparations, during which time cross-winds developed that cancelled the flight. The next night, the Cape experienced clear, very cold weather, down to the low 20s. Program managers realized these conditions might pose problems, but after engineers failed to discover any reasons to postpone the launch, technicians continued the countdown and pumped propellants into the external tank. Very early that morning—hours before dawn on January 28, 1986—an ice team inspected Pad 39B (this was to be the first Shuttle flight from this site) and found a significant accumulation. They returned after sunrise and on their advice the program managers

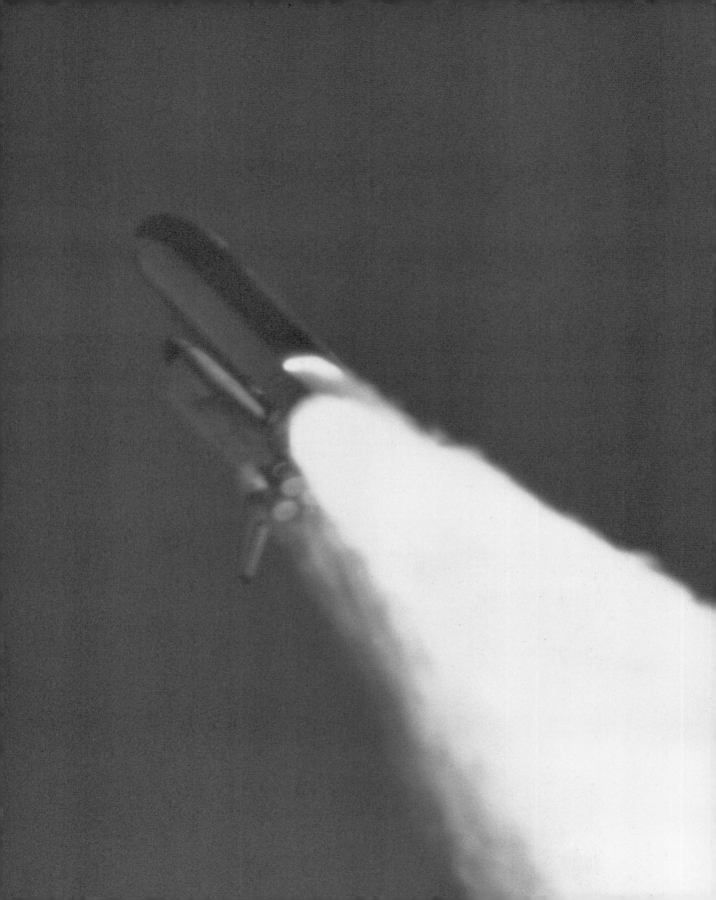

[Opposite] *This photograph depicts events fifty-nine seconds after launch of* Challenger. *At this point, the solid rocket booster is engulfed in flame. Six seconds later, the solid rocket booster's flame would break through the external tank's skin, causing the immense structure to collapse and fail.*

[Left] *A violent explosion followed the collapse of the external tank, as a ball of gas (fed by the external tank) shrouded the main engines and the solid rocket booster. At this point, the full horror of events dawned on the viewers at Kennedy, and on those watching on television.*

[Above] *The final destruction of* Challenger. *The force of the blast thrust the orbiter forward erratically and with such power that it broke up, hurtling its left wing, main engines, and crew cabin in all directions, and ending the lives of the crew. The orbiter's debris eventually fell into the Atlantic Ocean.*

Seated for breakfast the morning of lift-off, the crew of STS-33 included (left to right) mission specialist Ellison Onizuka, teacher in space Sharon Christa McAuliffe, pilot Michael Smith, commander Francis (Dick) Scobee, mission specialists Judith Resnik and Ronald McNair, and payload specialist Gregory Jarvis. The Challenger astronauts sat for an official portrait about two months earlier.

decided to wait until the ice melted. At twenty minutes before launch (already postponed for two hours) the team made a final check. At 11:38 Eastern Time the engines howled, and the stack rose. The ambient temperature was 15 degrees colder than any previous shuttle lift-off.

A sickening sequence of events followed. Had anyone looked for it, at the moment of launch (0.678 of a second into the sequence) they might have noticed a faint puff of smoke emanating from the field joint between the lower sections of the right solid rocket booster. This ball of vaporized matter suggested that the joint had been compromised somehow. From the same source, eight more puffs (with increasingly dark smoke due to the burning grease, insulation, and rubber at the joint) appeared during the following two and a half seconds, even as the massive Shuttle stack cleared the ground. At fifty-nine seconds, as the boosters increased thrust, the first sign of flame appeared. Little more than a second later, it became a plume visible without image enhancement.

Then the catastrophe occurred. As the pressure fell in the right booster, the flame grew in size. Guided by the aerodynamic slipstream, the finger of fire touched the surface of the external tank, at the same time affecting the strut holding the booster to the tank. Nearly sixty-five seconds after launch, the flame from the booster broke through the tank's exterior, causing a change in the plume as it mixed with hydrogen from the external tank. Almost instantly, the underside of the orbiter started to glow as it absorbed the heat of the external tank. Between seventy-two and seventy-three seconds after lifting off, the entire

STS experienced massive failure. First, the right booster became severed from the external tank, then the aft dome of the tank broke away, immense amounts of liquid hydrogen escaped, the booster collided with the bottom of the liquid oxygen tank in the intertank structure, and the mixing of hydrogen and oxygen in proximity of the flame resulted in an explosion of enormous power. *Challenger* broke into sections and fell away from the fireball, the forward fuselage visible as it plunged toward the Atlantic Ocean.

The seven crew members fatefully sealed in this part of the falling wreckage represented an extraordinary cross-section, but by no means an average group of Americans. *Challenger*'s commander, Francis R. "Dick" Scobee (born in 1939), enlisted in the U.S. Air Force in 1957 and after receiving a degree in aerospace engineering entered pilot training. After earning his wings in 1966, he flew combat in the Vietnam War and afterwards became a test pilot, flying forty-five types of aircraft and amassing 6500 hours in the cockpit. Dick Scobee entered the astronaut corps in 1978 and piloted the fifth mission of *Challenger* in April 1984.

Michael J. Smith served as the pilot of the doomed flight. Born in 1945 in Beaufort, North Carolina, he attended the U.S. Naval Academy and later earned a master's degree in aeronautical engineering from the Naval Postgraduate School. Smith flew A-6 Intruders aboard the USS *Kitty Hawk* during the conflict in Southeast Asia. As a Navy test pilot he experienced twenty-eight types of aircraft and accumulated

⋯⋯⟩ *Christa McAuliffe expanded the appeal of the U.S. space program to many, in part because of her refreshingly unpretentious attitude toward her assignment. In this picture, she meets her fellow crew members Michael Smith (left), Ronald McNair, and Dick Scobee during late summer 1985.*

[Opposite, top] *Because of all of
the publicity associated with her,
Christa McAuliffe also drew
attention to the increasing
numbers of women applying for
admission to the astronaut corps,
and receiving it. Images like this
one helped the cause. McAuliffe,
in flight suit (second from right)
walks with Barbara Morgan (left,
McAuliffe's backup on STS-33),
Michael Smith, and Dick Scobee
after flights in T-38 jet trainers.*

[Opposite, bottom row] *Christa
McAuliffe straps herself into a
seat (reserved for mission
specialists) behind Shuttle pilot
Michael Smith during a training
session at Johnson's Shuttle
mission simulator. During the
same day, McAuliffe took the
second (pilot's) seat on the flight
deck, where she learned about
some of the Shuttle flight systems
from Dick Scobee.*

[Below and left] *The images of
female astronauts training for
flights and going on missions
encouraged others. Here, Christa
McAuliffe (center) with her
backup Barbara Morgan
experience the exhilaration of
weightlessness aboard a diving
KC-135 tanker outfitted by NASA
for this purpose. In the picture on
the left, mission specialist Judith
Resnik, positioned at an orbiter
interdeck access hatchway, posed
for this picture with cameras
floating around her head.*

229

↑ *About five weeks after the* Challenger *disaster, members of the presidential commission investigating its causes met at the Kennedy Space Center to gather facts. Robert Holz (center), Dr. Sally Ride, and commission staffers arrive at the Launch Control Center.*

↓ *The next day (March 4, 1986) commission members pressed their investigation at the Johnson Space Center. Conferring in the executive conference room in JSC's Project Management Building are NASA Johnson deputy director Robert Goetz (far left) and deputy manager of the National Space Transportation System Richard Kohrs (seated). They are speaking to three Rogers Commission members: (left to right) Arthur Walker, Robert Rummel, and Joseph Sutter, Jr.*

4300 flying hours. STS-33 represented his first trip into space.

Judith Resnik of Akron, Ohio—one of the three mission specialists aboard the flight—earned a doctorate in electrical engineering from the University of Maryland in 1977. She entered the astronaut corps in 1978 (the first class to admit women) after several years in the department of neurophysiology at the National Institutes of Health. Dr. Resnik became the second American woman in space when she flew aboard the initial flight of *Discovery* in August and September 1984, contributing to biomedical experiments and to deploying three satellites.

Ronald E. McNair, the second mission specialist, grew up in Lake City, South Carolina, and despite the inequities of racial segregation, became valedictorian of his high school class. McNair received a doctorate in physics at the Massachusetts Institute of Technology in 1977. Recognized widely for his scholarly work on chemical and high pressure lasers, he worked in the optical physics department at Hughes Laboratory in Malibu, California. Like Reznik, he entered astronaut training in the historic class of 1978, which included the first three African-Americans. The second African-American in space, McNair flew aboard Challenger mission 11 in February 1984 and operated the Shuttle's Canadian-made maneuvering arm.

The third mission specialist—Ellison S. Onizuka of Hawaii—received bachelor's and master's degrees from the University of Colorado and entered the U.S. Air Force in 1970. After serving at the Sacramento Air Logistics Center as a flight test engineer and as a squadron flight

Richard P. Feynman

Born in Far Rockaway, New York, in May 1918 to immigrant parents from Minsk, Belarus, Richard Feynman showed an early love of mathematics, but also an equal bent towards practicality. Not surprisingly, in 1935 he entered the Massachusetts Institute of Technology. Put off by math professors who seemed to believe that the discipline existed for its own sake, he turned to electrical engineering, and then to physics. He enjoyed MIT, compiled a brilliant record in math and physics (but a mediocre one in history and literature), and then transferred to Princeton where he received the doctorate in physics in 1942. Between 1943 and 1945, Feynman—only in his mid-twenties—worked on the atomic bomb at Los Alamos and became chief of theoretical physics there. During the next five years he abandoned research and taught at Cornell University, but after moving west to Caltech in 1950, as professor of theoretical physics, he resumed an earlier interest in quantum mechanics. During the 1950s Feynman concentrated on quantum electrodynamics; he went on to win the Nobel Prize in 1965. A popular and highly original lecturer, Richard Feynman's contributions to the *Challenger* investigation owed nearly as much to his flamboyant manner of presentation, honed in the classroom, as to his undoubted brilliance. Ill-looking during his service on the Rogers Commission, Feynman nonetheless returned to teaching at Caltech afterwards. The following year he succumbed to abdominal cancer at the age of sixty-nine after an eight year struggle that may have had its origins in his nuclear research at Los Alamos.

test officer at the Air Force Flight Test Center, in January 1978 Onizuka joined the astronaut corps. Due to his USAF affiliation, he served on the first classified Defense Department Shuttle mission (STS-20) in January 1985 aboard *Discovery*.

The final two crew members of *Challenger*, both private citizens, had no ties to federal government service. Gregory B. Jarvis, a native of Detroit, Michigan, worked for Hughes Aircraft's Space and Communications Group in Los Angeles, California. Jarvis had earned the bachelor's and master's degrees in electrical engineering and won admission to the astronaut program in 1984 under Hughes sponsorship.

The last of the *Challenger* astronauts, Sharon Christa McAuliffe (born Christa Corrigan in 1948), had competed among more than 11,500 teachers for a spot in the Teacher in Space program, which she won in 1984. A New Englander raised in Framingham, Massachusetts, she attended Framingham State College where she majored in history. She subsequently taught American history in the Washington, D.C., area and earned a master's degree in school administration from Bowie State University in Maryland. McAuliffe, her husband Steve (an attorney), and their two small children returned to New England in 1978 and she applied to NASA while a teacher in Concord (New Hampshire) High School. Likeable, athletic, and unpretentious, McAuliffe appealed to the media and

↑ ⟶ *Challenger commission chair William Rogers and other members of the group visit Kennedy on March 7, 1986. Right, Rogers (far left) and Sally Ride are shown a portion of a solid rocket booster segment by center director Richard Smith. In the Vehicle Assembly Building, astronaut Robert Crippen calls their attention to the Shuttle* Discovery *tile installation.*

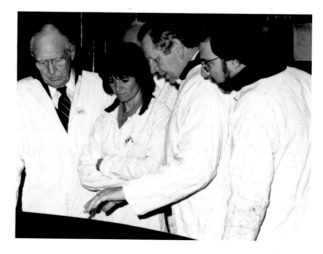

←⋯ ↑ Recovery teams found many sections of the Challenger *stack. Above, investigators display the forward skirt of the right solid rocket booster, a piece that transfers loads from the booster to the external tank and holds much of the booster's electrical and instrumentation systems. Left: Rather than destroying less critical parts of the wreckage, these men loaded boxes of debris into silos on Complex 31 at Cape Canaveral Air Force Station.*

↗ After months of combing the ocean floor with sonar and submarines, recovery teams completed their work and reconstruction crews started. They separated orbiter, solid booster, and external tank parts from one another. Then, in the Cape Canaveral Air Force Station impoundment areas, they used tape to lay a grid on the floor and placed the pieces in their correct relationship to each other. This photograph illustrates the debris collected from the orbiter portion of the wreckage.

gained a following among Americans as an astronaut with whom they could identify.

The same American public reacted with horror to the loss of *Challenger*. President Ronald Reagan led the nation in mourning. He also established an independent commission, chaired by former Secretary of State William P. Rogers, on February 6, 1986. They presented their findings to the president four months later to the day. Rogers and his distinguished panelists—including vice chair Neil Armstrong, Sally Ride, Air Force general Donald Kutyna, and test pilot Chuck Yeager—did not investigate the safety of the Space Shuttle as a whole, but instead confined themselves to the technical causes of the accident. But the members did attempt to determine whether other factors might reduce future risks.

Among this group of eminent individuals, one contributed the decisive insight about the cause of the *Challenger* explosion. During dramatic public hearings, Nobel laureate physicist Richard Feynman described in his homespun style how the disaster had originated with a gap in the O-ring that sealed the aft field joint of the right solid rocket booster, enabling gases to be exhausted into the atmosphere. Building on the Feynman thesis, Rogers and his group laid the final blame on faulty design that failed to take into account a number of important factors: cold ambient temperatures that rendered the O-ring more stiff and less capable of conforming to the contours required to make a tight seal; the physical dimensions of

the parts involved; the composition of the materials in use; the consequences of reusability; and the effects of dynamic loading on the joint.

But the criticism leveled by the Rogers Commission did not end there. The group posed sharp questions about two Shuttle contractors and NASA's reaction to their warnings. Evidently, engineers employed by Thiokol—the designer of the solid rocket boosters—had expressed concerns up their chain of command about the safety of a lift-off in the temperatures present on the morning of January 28, 1986. But fearful that they might tarnish their firm's reputation with managers at Marshall Space Flight Center who wanted to launch, Thiokol executives kept the bad news from NASA. Similarly, amongst themselves, Rockwell representatives raised questions (somewhat ambiguously) about icing, yet only told space agency officials that such conditions posed unknown risks. Ultimately, the commission faulted Marshall's leadership for concealing the flight's inadequacies from NASA headquarters, and for failing to exercise due caution amid warnings about the safety of *Challenger* and her crew.

The Space Shuttles remained grounded for two years and eight months as NASA program managers attempted to comply with the recommendations of the Rogers Commission. Luckily, James Fletcher agreed to return as Administrator, nine years after his first term as head of the space agency. Fletcher's appointment restored confidence inside and outside of NASA, enabling his staff to

↑ *Vice Admiral Richard Truly, who came from the astronaut corps, received the appointment as James Fletcher's successor from President George H.W. Bush on July 1, 1989. Truly impressed many as Deputy Associate Administrator for Spaceflight, a position in which he led the reconstruction of the Shuttle program and of the agency itself. He resigned as Administrator in February 1992.*

↘ *After thirty-two long months of inactivity, the Shuttle program roared back to life on the wings of* Discovery *and mission STS-26. The relatively small crew of five included (back row, left to right) mission specialists Mike Lounge, David Hilmers, and George Nelson, and (front row) pilot Richard Covey and commander Frederick Hauck.*

⋯⋯→ ⋯⋯→ *Images like that of pilot Dick Covey during STS-26R did much to calm fears that after the* Challenger *disaster and the long period of recovery, NASA might have lost its way permanently. Administrator James Fletcher also did much to restore confidence in the space agency.*

⋯⋯→ ⋯⋯→ ⋯⋯→ *Looking through Bay 3 of Kennedy's Vehicle Assembly Building at night, Shuttle* Discovery *is bathed in white xenon light as its transporter moves it to Pad 39B. The successful return to flight helped dispel doubts about NASA's capacity to provide the nation with heavy launch lift.*

concentrate on the immense task of returning the Shuttle to service. First, he asked former astronaut Richard Truly to assume direct control of the restoration of flight, which entailed a shake-up in program personnel. Fletcher also swung into action himself, ordering a replacement for *Challenger*, called *Endeavour*, overseeing the redesign of the solid rocket boosters, authorizing a better system of crew egress from the orbiter, imposing a new management structure on the institution, pouring money into Shuttle safety and reliability programs, and enacting a total redesign of Shuttle components to reduce the likelihood of failure. All of these labors culminated in the launch of *Discovery* on September 29, 1988, the twenty-sixth Space Shuttle lift-off. Despite a minor flaw in which the temperature in the crew compartment rose above 80° F (27° C), the flight lasted ninety-seven hours, involved sixty-three rotations around the Earth, and ended in the crew's safe return home.

Proving the solidity of Fletcher's achievement, by 2000 the Space Shuttle had flown for an additional twelve years and undergone seventy-five flights, certainly not without difficulties, but with no catastrophic failures like the one that befell *Challenger*. During this period, ten dedicated Department of Defense launches, nine dockings with the Russian *Mir* space station, and the first rendezvous with the International Space Station (in 1998) occurred.

Although the Space Shuttle failed to become the economical, on-time space truck desired by President Nixon—and despite its bewildering complexity—it still represented a gigantic technological stride towards practical access to space, and (for better or worse) engendered a more matter-of-fact attitude toward spaceflight on the part of the American people.

1. White House Press Secretary, "The White House, Statement of the President," March 7, 1970, quoted in Launius, *NASA: A History of the U.S. Civil Space Program*, pp. 219–20.

2. James R. Hansen, *Enchanted Rendezvous: John C. Houbolt and the Genesis of the Lunar-Orbit Rendezvous Concept*, NASA Monographs in Aerospace History Number 4, Washington, D.C. (NASA) 1995, p. 9.

3. Caspar W. Weinberger, Memorandum to the President, via George Schultz, "Future of NASA," August 12, 1971, quoted in Launius, *NASA: A History of the U.S. Civil Space Program*, pp. 223–24.

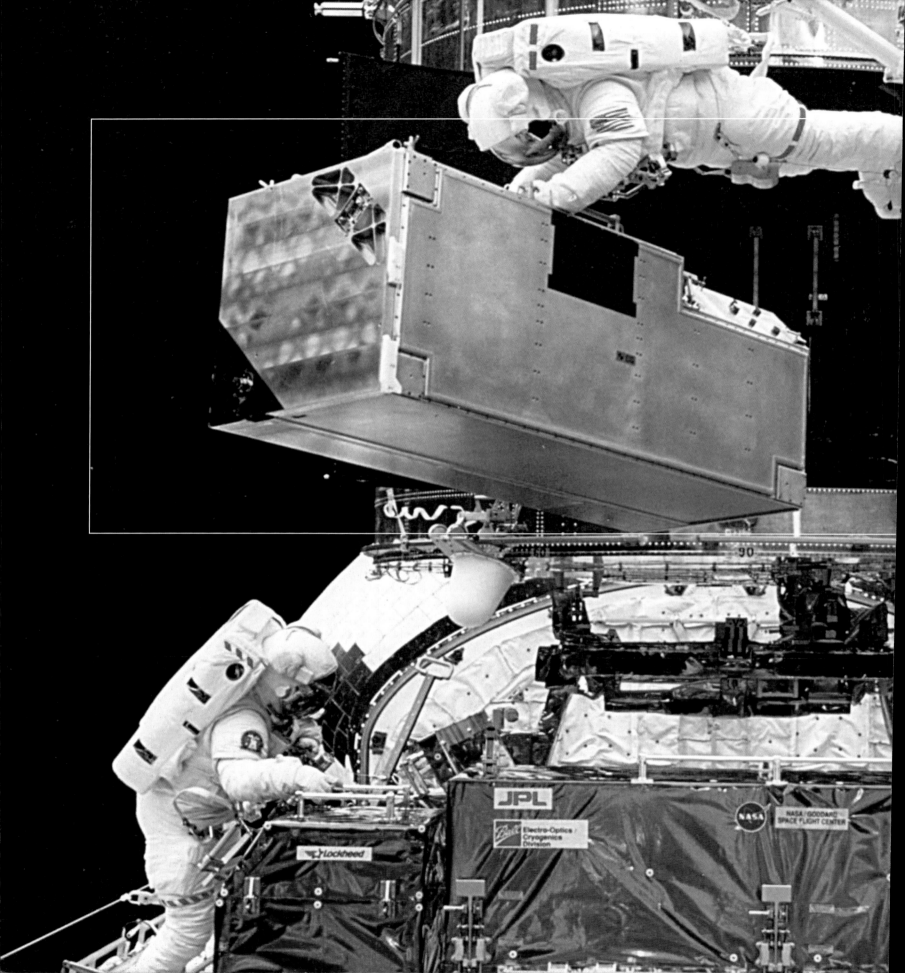

6

Round and Round
Spaceflight in the Twenty-first Century

Despite the many successes of the Space Shuttle after its return to flight, during the 1990s and into the new millennium the American space program encountered fundamental challenges to its survival. Perhaps the most severe problem stemmed from the absence of a set of coherent and long-term objectives. Indeed, cost cutting became perhaps the only consistent theme during this period. Throughout the decade and beyond, Congress steadily whittled the space agency's budget, in part because the legislators were growing tired of big ticket projects that brought uneven results.

In response, NASA's leaders pursued a full set of projects despite these limitations, adopting a management approach that emphasized quicker development of new vehicles, reduced outlays, and better quality. This unlikely formula resulted in some successes, but also in some famous failures.

Those who advocated the policy known as Faster, Better, Cheaper—most notably Daniel S. Goldin, the longest serving NASA Administrator (1992–2001)— believed that efficient and productive spaceflight demanded the simplification and miniaturization of space vehicles. Appointed by President George H.W. Bush to bring cost containment to the process of space exploration, Goldin championed smaller, less expensive satellites as an executive of the TRW Corporation. During his first six years at headquarters, transplanting the concept seemed to work: nine of ten spacecraft launched during this period succeeded. But then, starting in 1999, some stunning setbacks occurred, resulting in four failures to one success. The Mars Polar Lander typified the drastic reversal of fortune of the Goldin approach. Although outfitted with both a parachute and a descent engine, the vehicle's signal never materialized after the expected communications blackout that accompanied the lander's final approach to the Red Planet. True to the tight-fisted priorities of Faster, Better, Cheaper, designers had *dispensed with telemetry*, the only means by which controllers might have discovered what went wrong in time to fix it.

In the end, Dan Goldin tried to infuse NASA with a new spirit of optimism and purpose. But his ambitious efforts to reshape the space agency ended in mixed results, in part because he inherited some enormous projects with varied outcomes. One originated at the birth of the U.S. space program. During hearings before Congress in 1959, Fred L. Whipple—not only the director of the Smithsonian Institution's Astrophysical Observatory but also a man known to adults and children alike as a leading popularizer of the wonders of space—pressed the case for telescopes in the heavens. In 1968 the first incarnation of Whipple's dream circled the Earth. The Orbiting Astronomical Observatory-1 gazed at the universe through the media of ultraviolet, gamma ray, x-ray, and infrared radiation.

But even before the Orbiting Astronomical Observatory went into orbit, the National Academy of Sciences Space Science Board proposed in 1965 that NASA adapt the *Saturn V* to lift a much larger, diffraction limited telescope into space.[1] A few years later, some in the space science community envisioned a powerful eye in the cosmos capable of seeing to the far corners of the universe, sent aloft in the cargo bay of a spaceplane. Indeed, the Space Shuttle–"Big Telescope" pairing matured around 1971, and each project drew strength from the other. Shuttle planners embraced the celestial telescope because it clinched the case for reusable, heavy lift rocketry capable of propelling massive payloads into orbit and returning to

←···· *Daniel Goldin served as NASA Administrator for nine years and seven months, the longest serving NASA Administrator in history, under presidents George H.W. Bush, Bill Clinton, and George W. Bush. Tough-minded and controversial, Goldin imported business practices from his private sector experience and pushed for more affordable spaceflight, at the same time pursuing several high cost, marquee programs.*

Daniel S. Goldin

NASA Administrator Daniel S. Goldin (born in 1940) undertook perhaps the most ambitious reforms in the agency's history. Goldin grew up in a row house in New York City's South Bronx, in a family of few advantages. Motivated by his father and an uncle to study rockets and space, Goldin earned a bachelor's degree in mechanical engineering from the City College of New York. He then took a job with the Lewis (now Glenn) Research Center as an electric propulsion specialist during the height of the Apollo program. But rather than climb the career rungs at the space agency, by 1967 Goldin had decided to enter private industry and take a position with the California-based TRW Corporation, where he distinguished himself for his management of advanced (robotic) spacecraft. He rose to the rank of vice president and general manager of the firm's Space and Technology Group in Redondo Beach, California, where he worked on reconnaissance and communications satellites, as well as the Defense Department's Brilliant Pebbles initiative. His career at TRW taught him to value economical, lightweight, simpler space vehicles in contrast to the bigger, more complex machines often preferred by NASA. He got the chance to practice his ideas when President George H.W. Bush asked Administrator Richard Truly to step down and appointed Goldin in April 1992 to succeed him—and to tighten the agency's belt. Later on, Goldin continued to serve under presidents William J. Clinton and George W. Bush. Once in Washington, the new Administrator pressed the agency as none before him to cut costs, streamline operations, and produce cheaper spacecraft in greater numbers. Far from popular inside the agency, Goldin reduced the number of civil servant employees significantly, outsourced tasks (such as Space Shuttle support) hitherto undertaken by agency employees, and pared down its long-range spending plans by some $30,000,000,000. Despite some administrative and programmatic successes in the service of "Faster, Better, Cheaper," the two triumphs of Goldin's term—improving the astigmatism of the Hubble Telescope and recruiting the Russians as space station partners—proved to be two of the most complicated and costly projects ever conceived by NASA. Dan Goldin left the space agency in November 2001 (Sean O'Keefe succeeded him in December 2001) and went on to conduct robotics research at the Neurosciences Institute in La Jolla, California.

↑ ↗ *Goldin's international forays did not begin and end with the International Space Station. Despite his passion for frugality, the Cassini/Huygens mission to Saturn, like the massive space station, proved to be expensive. It also represented a true multinational alliance, resulting in a probe unlike any other: highly complicated, very heavy (12,600 pounds/5715 kg),* *and very costly ($3,300,000,000). Planning started in 1982, and after lifting off in October 1997 the probe reached the ringed planet in summer 2004, sending back breathtaking images. Cassini is expected to orbit Saturn for centuries. Pictured here, Cassini's upper equipment module and a view of Saturn taken by Cassini in February 2004.*

them routinely for maintenance, adjustment, and repair. For their part, the telescope advocates backed the shuttle as the most likely means of achieving their objectives. Inspired by this prospect, NASA engineers began to develop the pointing system for this revolutionary instrument, known subsequently to the world as the Hubble Space Telescope (HST).

Although design studies for the Hubble progressed in the early 1970s, the project lay fallow during this period due to tight budgets. Started anew in 1977 and given the firm go-ahead in 1978, the proposed system seemed like a marvel, capable of remarkable things. The size of a railroad tank car (43 feet/13 m long and 12 tons/ 12.1 tonnes in weight), its huge 94 inch (239 cm) primary mirror—perfect to micro inch tolerances on its surface— promised unparalleled views of stellar, galactic, and

↑ *The Orbiting Astronomical Observatory undergoing a shroud test in August 1965. A distant and humble relative of the Hubble Space Telescope, in 1968 it began circling the Earth, sending home images of the universe.*

extragalactic objects as far as 15,000,000,000 light years distant.

Soon, however, the mighty telescope ran into trouble. By early 1983, NASA Administrator James Beggs was facing a crisis. Behind schedule and working out significantly more expensive than anticipated, the Hubble needed to be bailed out by Congress. The nation's legislators did so, but only on the conditions of no more overruns and a shake-up in the program management. The glare of publicity seemed to bring the HST out of the shadows. Goddard Space Flight Center assumed responsibility both for the program and for the contributions of the Space Telescope Science Institute, a university consortium located at Johns Hopkins University. The European Space Agency signed agreements with NASA to enable astronomers from overseas to gain access

to the instrument. Finally, just as the horrible news of *Challenger* became known, engineers and technicians assembling the machine at the Lockheed plant in Sunnyvale, California, exposed the nearly completed space telescope to a battery of ground tests.

Once again, the fates of the Shuttle and the telescope intersected. During the long delay prior to the resumption of STS flight, the Hubble remained in storage in California, incapable of being boosted out of the atmosphere by any other means. Conversely, once the Shuttle returned to service, the media associated the Hubble Space Telescope (by this time a project costing more than $2,000,000,000) with the Shuttle's full return to operations.

NASA mission planners chose Flight 35 (STS-31R) for the telescope launch, perhaps the most important non-human payload ever carried aloft by the Space

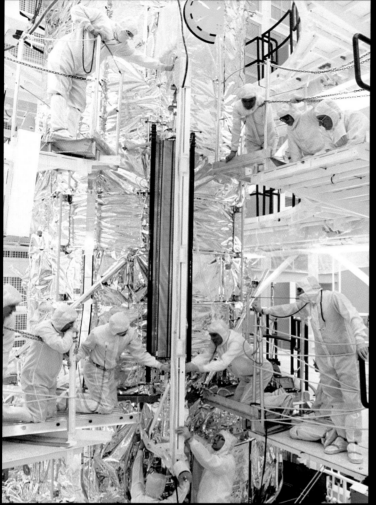

[Above and opposite] *Three views of the mighty Hubble Telescope in different stages of preparation: the telescope itself undergoing assembly in the cleanroom (above right); the wide field/planetary camera being readied for installation (far right); and a fit check of the Solar Array flight article (above left). Marshall Space Flight Center assumed primary responsibility for Hubble's design and fabrication, contracting with Perkins Elmer for the optics and Lockheed Martin Missile and Space for the spacecraft systems and the final testing of the telescope.*

[Right] *Because of HST's size, no launch vehicle but the Space Shuttle had the capacity to lift the Hubble Telescope into orbit. As a consequence, the* Challenger *tragedy and the long wait for return to flight postponed its deployment. During this period, the telescope remained at the Lockheed Missile and Space plant in Sunnyvale, California. Here, technicians complete the assembly and testing of the spacecraft.*

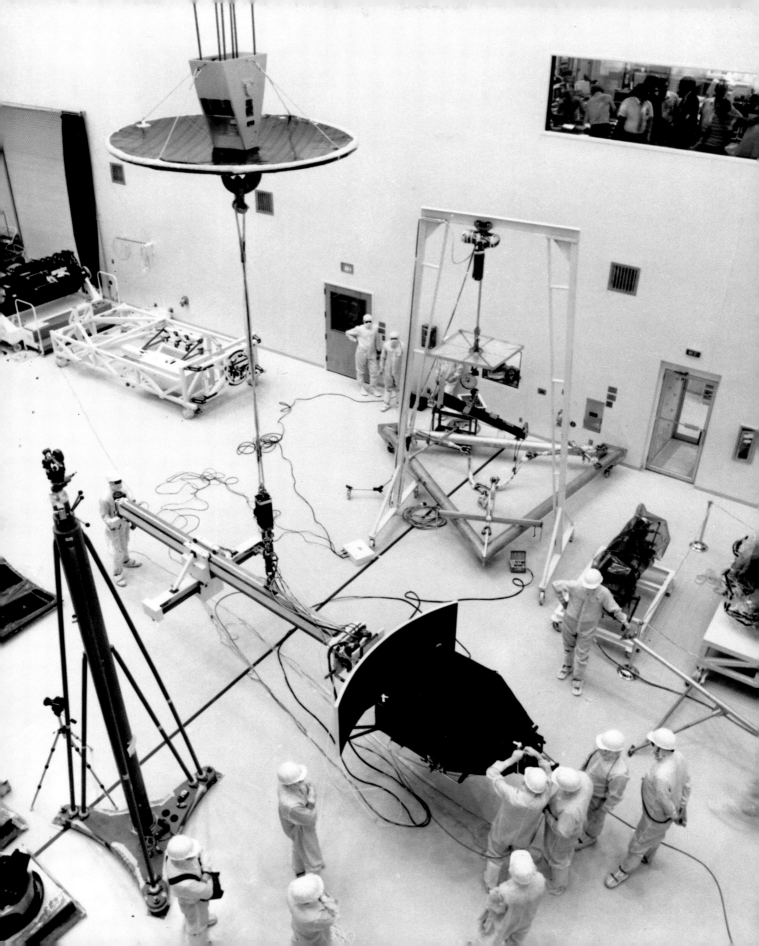

Edwin P. Hubble

Edwin Powell Hubble (1889–1953), the namesake of the Hubble Space Telescope, transformed the prevailing wisdom about the structure and dynamics of the universe. The son of John Hubble, an insurance man from Missouri, and Virginia Lee (James) of Virginia City, Nevada, Edwin grew up in Wheaton, Illinois, a suburb of Chicago. He won a half scholarship to the University of Chicago, served as a laboratory assistant for future Nobel laureate and Caltech president Robert Andrews Millikan, and earned undergraduate degrees in math and astronomy. Hubble then took a professional and personal detour by accepting a Rhodes scholarship to Queen's College, Oxford, where he studied law. But after a year of uninspired legal practice he returned to his first love, re-entered the University of Chicago, and received a doctorate in astronomy. There he fell under the influence of fellow Chicagoan George Ellery Hale, one of the great American astronomers of the preceding generation. Hale, director of the Mount Wilson Observatory in Southern California, invited him in 1917 to come to Pasadena, California, and join his team. Two years later, at the age of thirty, he began his career at Mount Wilson. Hubble also got married, relatively late, to Grace Burke in February 1924. During the 1920s and 1930s, using the Hooker Telescope—the world's biggest, a 98 inch (249 cm) reflector perched high above the city of Los Angeles—he made his three great contributions to astronomy. First, he showed that the Milky Way, rather than being the universe's single, great galaxy, actually represented one of many. Second, he arrived at the startling observational conclusion—posited theoretically by Albert Einstein in 1917—that the universe existed in a state of outwards expansion. Finally, Hubble established a classification system for the galaxies based on content, shape, brightness, and distance. During World War II Edwin Hubble served his country by pursuing research at the Army's Aberdeen Proving Grounds in Maryland, for which he received the Medal of Merit in 1946. He returned to California after the war, resumed his scientific investigations, and became a moving force in the creation of the mighty 200 inch (508 cm) Hale Telescope erected on Mount Palomar, California. Hubble died unexpectedly from a cerebral thrombosis in September 1953.

Transportation System. Originally they selected April 18, 1990, but after a readiness review, decided on April 10 (the first time a Shuttle launch had been attempted ahead of schedule). However, four minutes before lift-off, a malfunctioning valve caused the flight to be postponed, resulting in a two week delay in order to recharge the HST's batteries. Accordingly, the nation watched anxiously as the orbiter Discovery—its cargo bay stuffed with the hulking instrument—lifted off from Kennedy Space Center's Pad B on April 24, 1990.

To the surprise of many, the release of the Hubble seemed almost an anticlimax, especially after the years of painstaking rehearsals. The spacecraft was conceived to be fully repairable in orbit, and its designers had taken every precaution to render it easy to fix and maintain. Dozens of yellow handholds and sockets for foot restraints dotted the telescope's exterior sheathing. Hinged doors opened to reveal parts that were removable by simple tools. Moreover, in preparation for the launch of Hubble, astronauts Kathy Sullivan and Bruce McCandless (both aboard STS-31R) had practiced emergency maneuvers time and again in the water tanks at Marshall and Johnson. Indeed, McCandless had devoted much of the previous twenty years to developing and perfecting the tools and techniques necessary to keep the telescope functioning during its predicted fifteen year lifespan.

Then, about twenty-four hours after launch, as the five person crew aboard Discovery and millions on Earth held their breath, astronaut and astronomer Steven Hawley pried the HST out of the payload bay using the Canadian-made Remote Manipulator System. More commonly known as the Shuttle arm, this 50 foot (15 m) long, six-jointed appendage—capable of hoisting up to 65,000 pounds (29,00 kg)—served Hawley well. As the freed Hubble dangled from the mechanical limb, controllers on Earth signaled it to open its fan-like solar arrays and to deploy its antennas. The only uncertainty occurred when one of the arrays stuck. Sullivan and McCandless suited up for a possible EVA, but they stayed in the airlock as the controllers fixed the problem. Hawley then released the behemoth, which pointed itself towards the Sun. As Discovery receded into blackness, Hubble received commands to actuate its systems and awaken itself.

THE RELEASE OF THE HUBBLE SEEMED ALMOST AN ANTICLIMAX, ESPECIALLY AFTER ALL THE YEARS OF PAINSTAKING REHEARSALS.

[Above] *Its cargo bay packed with the Hubble Telescope,* Discovery *left Pad 39A on April 24, 1990. Having undergone intense planning and rehearsals, Hubble's five person crew made the actual release seem oddly routine.*

[Far left] *The delicate role of cajoling the Hubble out of the cargo bay fell to Steven Hawley, astronomer and mission specialist on STS-41D. His careful maneuvering of the Shuttle arm finally liberated the telescope, but the elation did not last long.*

[Left] *Kathryn Sullivan (top), a member of the first class of female astronauts and veteran of STS-41D (October 1984) practiced exhaustively for the eventuality that something might go wrong during the initial deployment of the telescope. In this picture (taken during the 1984 voyage) she uses binoculars to get a good view of Earth. Astronaut Bruce McCandless (below) probably knew more about the exterior of the Hubble Space Telescope than anyone else. Indeed, he devised many of the practices relating to the vehicle's maintenance and repair.*

↑⟶ *After the initial disappointment about the blurred images from the Hubble, NASA officials decided on a rescue mission. During 1993, while the technical preparations advanced, Shuttle astronauts practiced in space the techniques and maneuvers required to fix the space telescope. STS-57 astronauts David Low and Peter Wisoff (opposite) simulated the repair mission using the Shuttle arm in June 1993. More simulations followed in September of the same year (above).*

In contrast to the surprising ease of the deployment, once the telescope flew independently its troubles mounted rapidly. The motions of its many parts prompted the HST system to switch into safe mode, causing its computers to shut down repeatedly. Sensors used for guidance kept locking onto stars. The spacecraft oscillated, confounding its precise pointing system. Once the project engineers had surmounted these and other initial peculiarities of the HST, in late May they finally looked at the first images. Joy, then despair washed over the world's astronomical community. Initially, the pictures looked perfect, clear and sharp against the neighboring darkness. But in less than a month, a disastrous flaw became obvious. In focusing the system's mirrors on these objects (especially stars) the Hubble failed to focus all of the incoming light, resulting in photographs that looked like pinpoints encircled by rings or haloes.

An investigation in July uncovered the unthinkable. The computer-controlled machines that had ground and polished the primary mirror to near perfection had been

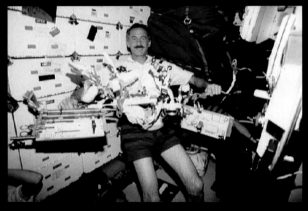

[Top] *In one of the most daring events in the history of spaceflight, in December 1993 the crew of STS-61 attempted to repair the multi-billion dollar Hubble Space Telescope while in orbit. They included, seated left to right, pilot Kenneth Bowersox and mission specialists Kathryn Thornton, F. Story Musgrave, and Claude Nicollier; and standing left to right, commander Richard Covey and mission specialists Jeffrey Hoffman and Thomas Akers.*

[Above] *Three of the astronauts who restored Hubble's vision: (left to right) Jeffrey Hoffman during the actual repair flight, in the mid-deck area, displaying some of the tools for the mission; Story Musgrave, shown during training in 1989; and Thomas Akers during training in 1990.*

[Right] *During the Hubble repair mission aboard the Shuttle* Endeavour, *Story Musgrave, tethered to the Shuttle arm, is being raised to the top of the huge telescope in order to install protective covers on the machine's magnetometers. Jeffrey Hoffman (in the cargo bay) attends to other chores.*

[Opposite] *Astronaut Jeffrey Hoffman (in the center of the photograph), anchored by his feet to the Shuttle arm, prepares to participate in the replacement of the telescope's wide field/ planetary camera. Hoffman and Musgrave (lower left corner) extract the replacement camera from the scientific instrument protective enclosure.*

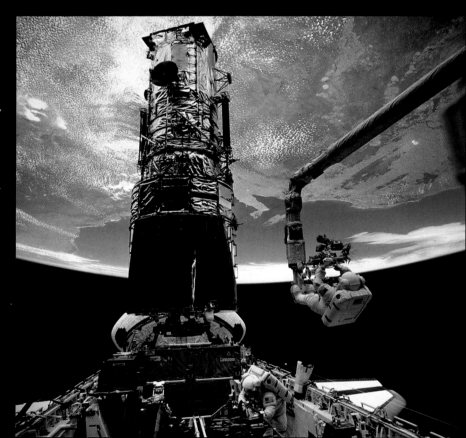

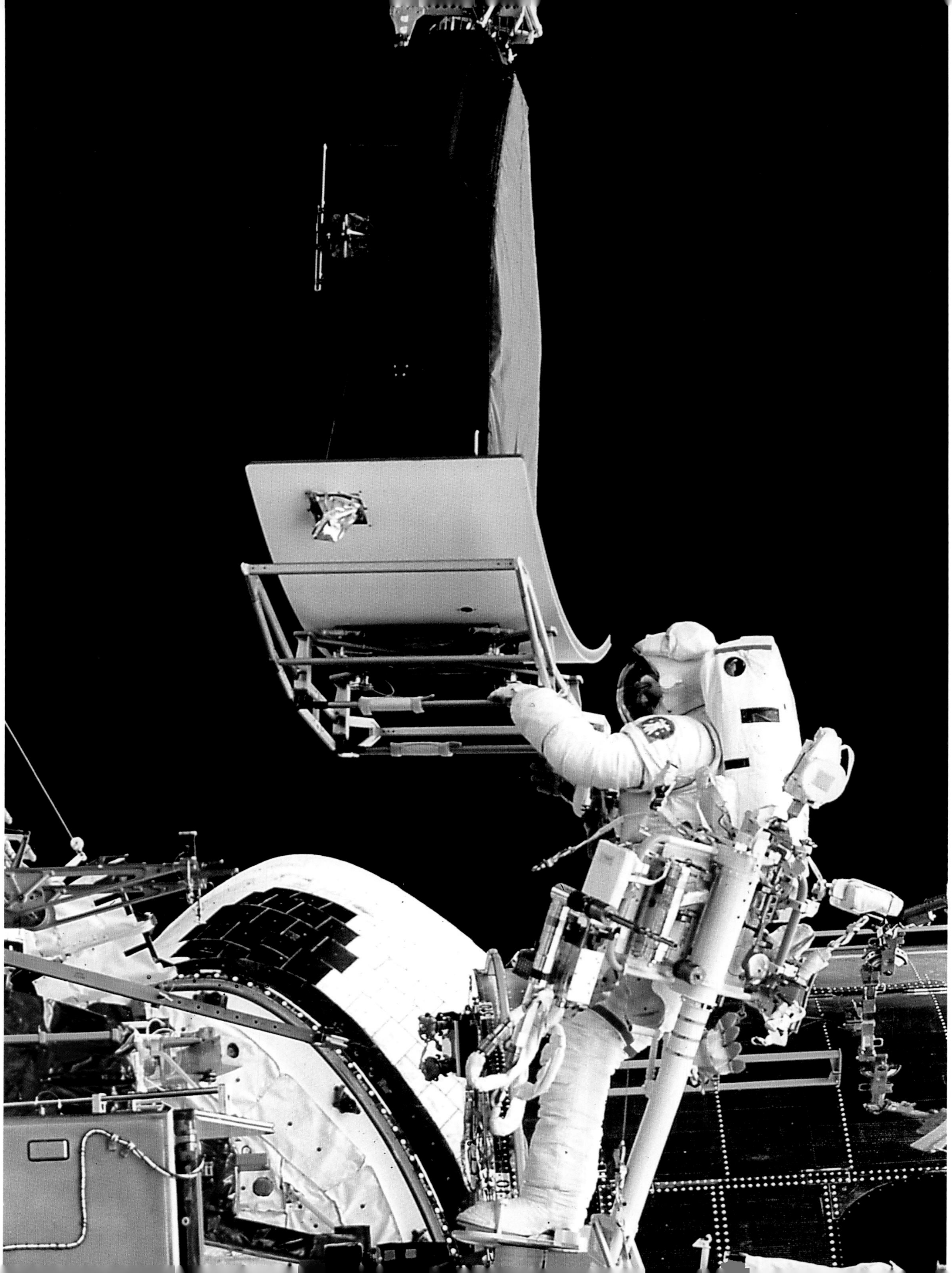

← Jeffrey Hoffman grapples with the original wide field/planetary camera, replaced by the new wide field/planetary camera in Endeavour's cargo bay. This mission did not end the servicing missions to Hubble; many followed in subsequent years.

↑ Kathy Thornton (above), moored to the Shuttle arm, lifts the Corrective Optics Space Telescope Axial Replacement (COSTAR) out of the cargo bay just before its installation on the Hubble. Thomas Akers, lower left, assists Thornton in the process.

informed by incorrect specifications. Apparently, the automated measurement checks had gone awry due to a faulty test set-up. But the headaches did not end there. Researchers learned that the telescope's strange oscillations, which made it so hard to point and to fix on objects, resulted from the heating and cooling of the solar panels as they passed into and out of the Earth's shadow. Additionally, one of the spacecraft's gyroscopes ceased to function, one of three fine guidance sensors had to be switched off, and the electronics package for one of the solar panels stopped working. The Hubble Space Telescope continued to return images of importance, but a feeling of bitter disappointment gripped the project.

Hard-driving Dan Goldin knew that the agency needed to extricate itself from the HST situation, not only to recover the promised technical capability but also to staunch the highly adverse publicity. Television host David Letterman, for example, lampooned HST on his show in

May 1990 with his "Top Ten Hubble Telescope Excuses," number seven of which said, "See if you can think straight after twelve days of drinking Tang."

Realizing that some members of Congress might take the opportunity of a lame Hubble either to starve or to end the program, NASA engineers lost no time in transforming the first HST servicing flight, scheduled for 1993, into a do-or-die repair mission. One piece of good news buoyed the Hubble scientists and engineers at Goddard and at the Space Telescope Institute at Hopkins. Although shaped incorrectly, the primary mirror had been ground wrong uniformly, thus making it possible to correct its contours just as an optometrist might compensate astigmatism with a corrective lens. By scouring the mirror's manufacturing records, the researchers concluded that the telescope's main mirror needed to be raised roughly two micrometers, or one-fortieth of the thickness of a human hair. Still, the

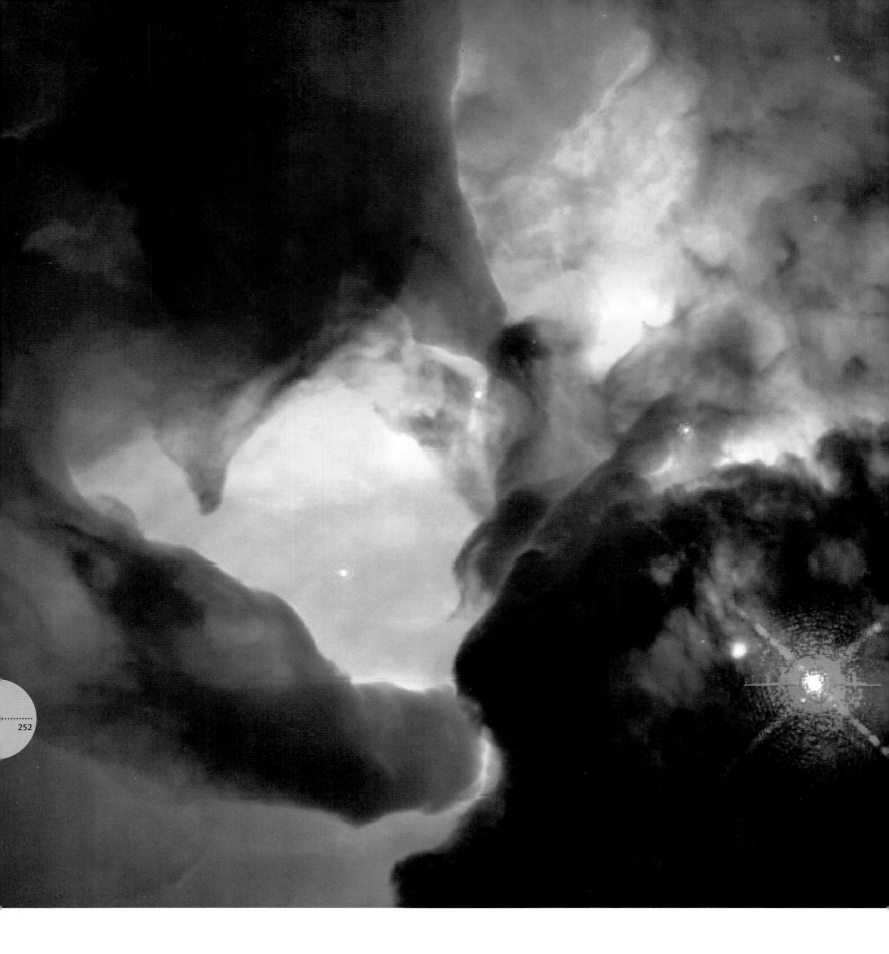

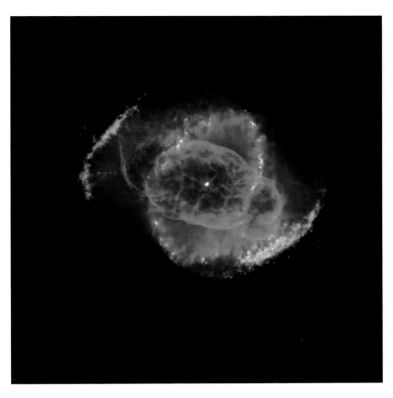

The images the Hubble Space Telescope sent home after the late 1993 repairs mesmerized the world and thrilled astronomers. Such views as the Cat's Eye Nebula in the Constellation Draco (above), the giant twisters in the Lagoon Nebula (left), and a Milky Way globular cluster of ancient stars illustrate the kind of pictures that shed light on the origins of the universe.

likelihood of achieving this objective in orbit, as well as fixing other, unrelated Hubble problems, seemed dubious. For instance, the remedy for a recurring oscillation in the spacecraft's flight motions did not lie in a software solution; the Hubble's on-board memory lacked the capacity for a long-distance repair. Hands-on computer replacement appeared to be the only answer.

After three and a half years of intensive preliminaries, the seven astronauts of the Hubble servicing flight readied themselves for launch aboard mission STS-61. They faced the most sustained, the most physically demanding, and perhaps the most complicated of space walks ever attempted, spanning ten days of flight and five separate EVAs. A scare just before lift-off threatened to wreak havoc on the mission. Construction sandblasting on Kennedy Pad 39A had left grit in the payload changeout room, where the telescope had been stored prior to being put aboard. Luckily, the HST internal package had been spared contamination because of its tight wrapping. The launch moved to Pad 39B, which glowed like a hearth as *Endeavour* sliced through the pre-dawn darkness and roared into the heavens on December 2, 1993. Two days later, Commander Richard Covey maneuvered the orbiter close to the Hubble, after which Claude Nicollier of the European Space Agency captured it with the Shuttle arm and tethered it to the aft payload bay.

Then the hard work began. Three main tasks lay ahead,

any one of which presented a formidable challenge. The astronauts needed to replace the solar arrays and their electronics; remove the old wide field/planetary camera and substitute an improved one; and install the corrective optics. Over the next eight hours, two of the four crew members with EVA experience (Story Musgrave and Jeffrey Hoffman) left the airlock and began work on the immense instrument. To conduct the repairs they counted on more than two hundred individual tools weighing an aggregate 14,400 pounds (6500 kg).

First, Musgrave and Hoffman succeeded in putting new rate sensing units into place. The next day, the other *Endeavour* EVA veterans, Kathy Thornton and Thomas Akers, spent more than six and a half hours removing the existing solar arrays and installing new ones. On December 6 Musgrave and Hoffman again left the safety of the orbiter, this time for nearly seven hours, to remove the 620 pound (280 kg) wide field/planetary camera and to equip the telescope with an updated version. Thornton and Akers returned the following day and devoted nearly seven hours to the all-important, 647 pound (293 kg) Corrective Optics Space Telescope Axial Replacement (COSTAR) module, as well as a new DF-224 computer processor. Finally, on December 8, Musgrave and Hoffman replaced the solar panels' electronic drive devices.

The results of the Hubble repair mission went on to become the stuff of legend. Millions each day now access the internet to witness the beauty and splendor reflected in Hubble's eye, trained on the far corners of the universe. Its restoration represented one of the finest technological achievements of the space program. But for sheer size, cost, and political complexity, the International Space Station (ISS) stands alone in the annals of air and space history.

In fact, the space station concept long predated the ISS. The early twentieth-century proponents of planetary exploration regarded large orbiting structures as indispensable staging areas, from which distant and ambitious journeys might be launched. Indeed, like the garrison among the European settlers of North America— or the base camp among adventurers in the polar regions—the space station recalled a familiar analog: the fortress in the wilderness.

During the 1940s and 1950s, a cultural cocktail mixing a faith in technology, the lure of high speed and hypersonic flight, and traditional American attitudes toward the wild coalesced in a fascination with permanent habitations in the cosmos. Space popularizer Willy Ley wrote in the 1944 edition of *Rockets* that exploration of the heavens depended on space stations as launch points for lunar and planetary missions. Space scientist Wernher von Braun's essay "Crossing the Last Frontier," published in 1952, heralded the dawn of the space age and the importance of big space stations. Perhaps most of all, artist Chesley Bonestell, in a series of sensational paintings published in *Collier's* magazine in March 1952, illustrated for the layperson von Braun's poetic vision: a

magnificent, serene, orbiting wheel 250 feet (76 m) across with a hub, spokes, and a thick outer rim to house and nurture its inhabitants.

In its modern incarnation, the concept of an orbiting, long-term home for space explorers originated as early as 1969. During that year, NASA engineers contacted their counterparts in industry and asked them to submit initial designs for an ambitious space station, capable of sheltering fifty to a hundred spacefarers. But this initiative came to naught, and real progress occurred only with the presidential administration of Ronald Reagan. Momentum for it gathered (not coincidentally) with the first flights of the Shuttle *Columbia*, the vehicle best suited to lifting the massive sections of the space station out of the atmosphere.

From the start of the Shuttle's operational career in 1981, NASA officials and space advocates urged the White House to give the space station center stage in the epic of human spaceflight, advocating it as an outpost on the way to the Moon and the other planets. NASA Administrator James Beggs, no doubt alert to Reagan's Cold War proclivities, also presented the project to the president as an opportunity to throw down the gauntlet to the U.S.S.R. Despite some sharp opposition among staffers concerned about deficits, the president announced his support, but with less flourish than the space constituency might have liked.

In January 1984, during his State of the Union address, Reagan directed NASA to initiate the space station and to

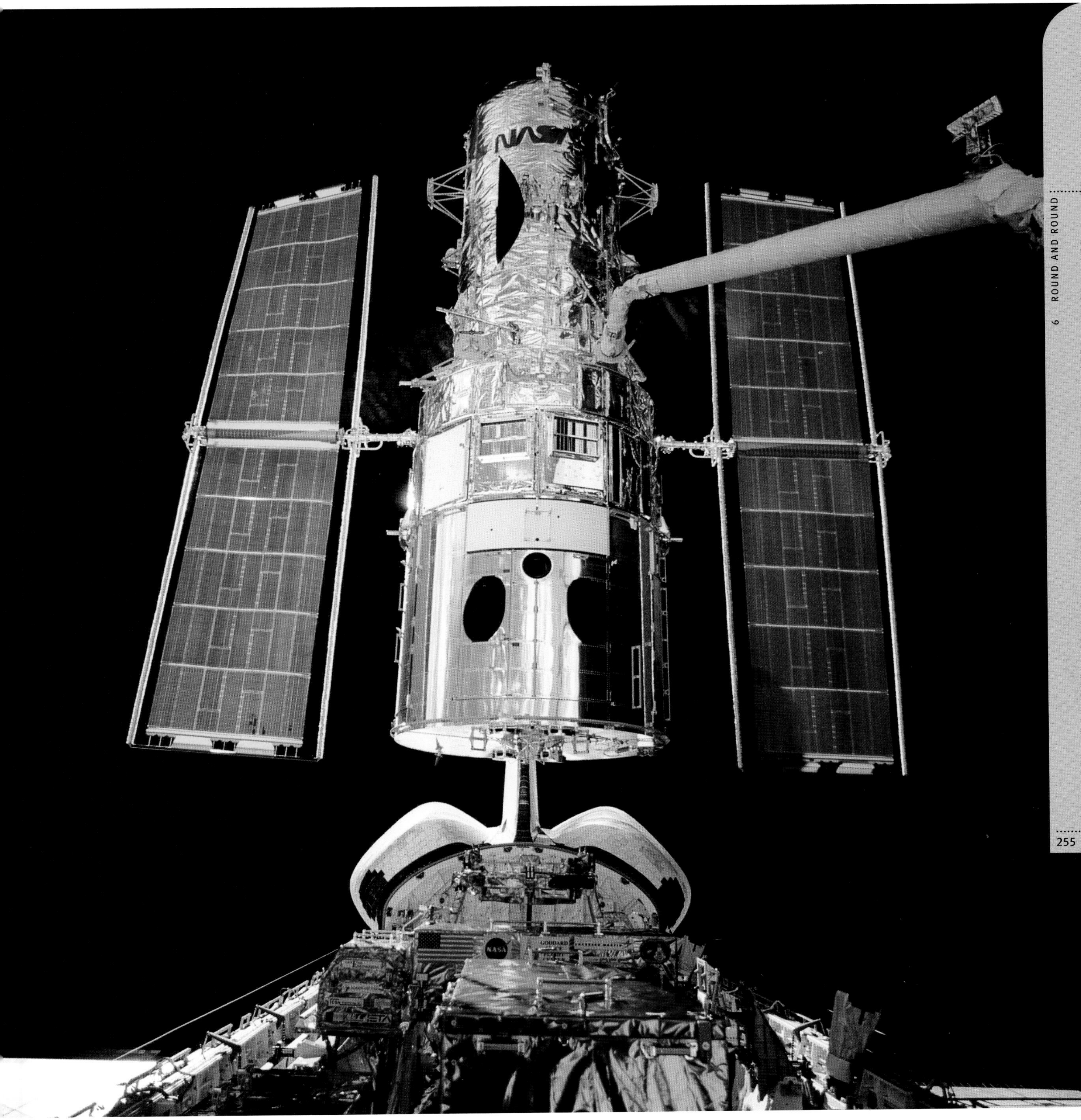

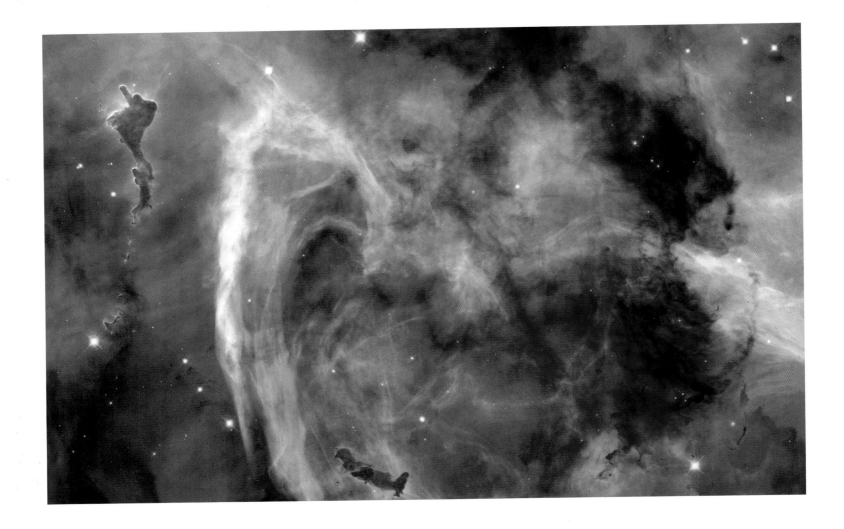

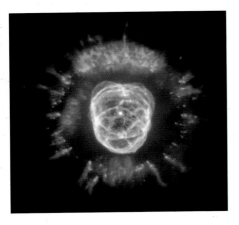

←--- ↑ ---→ *Once the Hubble's eye reopened, it continued to surprise and amaze. The first image it captured came from the Constellation Gemini, a cluster called the Eskimo Nebula (left); then it turned its lens on Carina Keyhole Nebula (above), and later in 2000 on the Glowing Eye Nebula in the Constellation Aquila (opposite).*

MILLIONS EACH DAY NOW ACCESS THE INTERNET TO WITNESS THE BEAUTY AND SPLENDOR REFLECTED IN HUBBLE'S EYE, TRAINED ON THE FAR CORNERS OF THE UNIVERSE.

complete it in ten years, a deadline reminiscent of the one President Kennedy had imposed on Apollo. But unlike Kennedy, who elevated the lunar landings into a cause, Reagan thought of the space station in more limited terms. Rather than a launch pad for bold forays into the heavens, he envisioned it as a research platform:

A space station will permit quantum leaps in our research in science, communications, in metals, and in lifesaving medicines which could be manufactured only in space. We want our friends to help us meet these challenges and share in their benefits. NASA will invite other countries to participate so we can strengthen peace, build prosperity, and expand freedom for all who share our goals.[2]

There were in fact some well-worn precedents to President Reagan's conception of the space station as an international partnership. Hugh L. Dryden, the first Deputy Administrator of NASA, had negotiated successfully with Soviet scientist Anatoliy A. Blagonravov for limited exchanges of meteorological, communications, and magnetic field data during the early space program. Many other scientific collaborations followed. During the

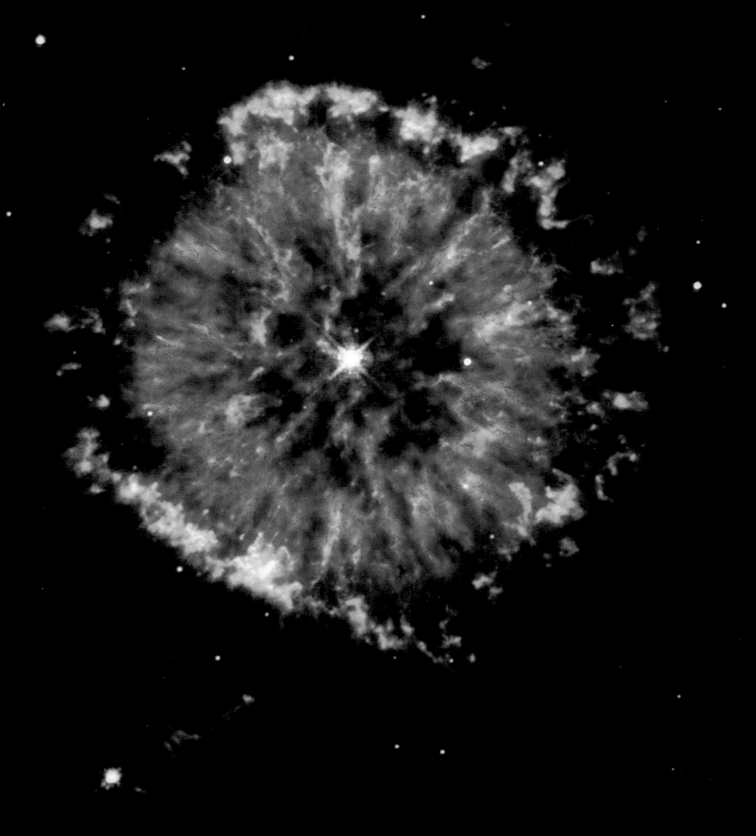

mid-1970s, the Apollo–Soyuz Test Project offered the first opportunity for astronauts and cosmonauts to join hands in the cosmos. The time-honored *Soyuz* docked with the equally venerable *Apollo* spacecraft in July 1975 and the crews migrated through the NASA-designed universal module to greet one another and to conduct experiments. During the following decade, scientists of the European Space Agency persuaded NASA officials to mount *Spacelab*, a cluster of miniaturized instruments, on the Shuttle cargo bay. Between 1983 and 1986 four Shuttle flights carried Spacelab, enabling data to be collected from its unique suite of tools.

The *Skylab* Orbital Workshop offered the closest precursor to the president's project. An astronomical platform in space, this 100 ton (101.6 tonne), relatively low cost orbiting laboratory (fashioned from the upper stage of the massive *Saturn V* rocket) became America's first space station after it lifted off from Kennedy Space Center in May 1973. It circled the world unoccupied for two weeks until a crew of three followed in an Apollo capsule, launched from a *Saturn IB*. The astronauts then rendezvoused with and occupied *Skylab*. Two more crews followed. Among the payloads of these missions, the Apollo Telescope Mount assumed the greatest prominence. A full-sized solar observatory, it was trained permanently on the Sun. The astronauts set it to target specific areas, such as flares (in order to measure the amount of radiation they generated). Significant for future missions, the astronauts learned to repair the instrument in flight, enabling it to operate without interruption during the six months of *Skylab*'s travels. Each mission lasted longer than the one before. *Skylab 2* stayed aloft from

May 25 to June 22, 1973, *Skylab 3* from July 28 to September 25, and *Skylab 4* from November 16 to February 8, 1974. Collectively, *Skylab* and its nine voyagers logged more than 171 days in orbit.

Still, scientists and engineers could glean only so much from a makeshift spacecraft designed to test limited endurance, as compared to a long-duration vehicle accommodating long-term residents. As a consequence, in 1985 NASA officials unveiled a proposal that answered President Reagan's space station declaration of the year before. Counting on the forces of gravity that the station itself generated, the designers of this vehicle abandoned the classic pinwheel shape proposed by the early space enthusiasts and instead conceived of a spaceship that looked as though it had sprung from a plumber's tool box. It consisted of experimental labs and astronaut living spaces close to the center of mass, dual vertical keels holding experiments and servicing equipment, and a boom dedicated to the solar power apparatus. Its creators called it Space Station *Freedom*.

But NASA Administrator James Beggs faced a formidable challenge bringing this creation to life. The public expected grandeur from a space station because of the images in the popular imagination fed by many fictional antecedents. In a sense, the NASA prototype fulfilled this requirement. It measured a stately 500 feet (152 m) in length and 360 feet (110 m) in height (although it housed no more than eight astronauts in a small cabin). Beggs decided to present *Freedom* to Congress as a multi-purpose spaceship, capable of serving as a staging point for voyages into the solar system, a laboratory, an observation post, a repair shop for broken satellites, and project—ranking with some of the greatest feats in the annals of human engineering—needed to be affordable to win and sustain congressional backing. Consequently, Beggs and NASA lured Canada, the European Space Agency, and Japan into a partnership and pledged to build the station for $8,000,000,000. This optimistic estimate—based on the reasonable expectation of incremental station expansion over time—was to bring the project untold woe during its long development.

Reversals piled up quickly, and in short order the dream of a massive, multi-purpose station evaporated. Within a year of the project's unveiling, budget realities resulted in the elimination of the vertical keels and with them, the capacity for on-orbit fabrication, assembly, repair, and maintenance, the core of the traditional concept of a stairway to the heavens. Only the laboratory remained. *Freedom* then underwent a long and painful period of contraction in the wake of the *Challenger* accident and the Shuttle's long hiatus from flight. What remained in 1991 bore little resemblance to the grand vision of the past. Reduced to a crew of four, who were confined to smaller quarters and laboratories, the missions became limited to microgravity and life science research. Upon taking office in 1993, President Bill Clinton's administration further scaled back the station's volume and number of inhabitants, and gave it the more modest name *Alpha*.

The saga of the space station then took a detour. NASA Administrator Daniel Goldin persuaded his Russian

---> *Responsible for the space station decision, President Ronald Reagan also lent his powerful backing to the space agency during the difficult period after the* Challenger *accident. Here, Reagan (left) presented several awards at the White House in May 1981 to (left to right): astronaut John Young (both the Congressional Space Medal of Honor and the NASA Distinguished Service Medal); Robert Crippen (the NASA Distinguished Service Medal); and Dr. Alan Lovelace (the President's Citizens Medal). Vice President George Bush (far right) looks on.*

Although Ronald Wilson Reagan (1911–2004), the fort president of the United States, did not embark on a h ambitious space program, he did govern during an important period in the history of spaceflight. He was born on February 6 in the town of Tampico, Illinois, to and Nelle Reagan. Ronald Reagan graduated from Eur College, a small Christian school, in 1932 with an aver academic record but success as a football player and actor. He took a job at WOC radio in Davenport, Iowa, in 1937 changed direction when he signed with the Warner Brothers studios as a contract player. This eve launched him as a film actor, a career in which he appeared in more than fifty motion pictures. Meanwh he took an increasingly active role in union activity, becoming president of the Screen Actors Guild in 1947 During the 1950s, Reagan's politics evolved gradually towards Conservatism. Reagan switched his allegianc from Democrats to Republicans in 1962. After two terr as Governor of California, Reagan decided to challeng Gerald Ford as the Republican party's nominee for president. Ford won narrowly, but lost to Jimmy Carter the election. In the 1980 presidential race, the sixty-ni year old Reagan won an easy victory over President Carter. The Reagan presidency spent heavily on defen in the end perhaps driving the Soviet Union into penu and collapse as it attempted unsuccessfully to keep p Reagan's space policy lacked the same zeal, but still witnessed some important events. The Space Shuttle began operational status three months after his first inauguration as president, and became the standard bearer of the U.S. space program in subsequent years The space station took shape at the start of Reagan's second term. Most importantly, as president during th *Challenger* accident in 1986, Reagan comforted the grieving families and the nation, appointed a distinguished panel to investigate the incident, and persuaded James Fletcher to return as NASA Administrator during the long and difficult period in w the shuttles remained grounded. Just before the end o his term in the White House, he celebrated the return flight of the American space program with the launch Discovery in September 1988. Afflicted with Alzheimer disease, President Reagan (the second longest-lived U president in history, after Gerald R. Ford) died in Los Angeles, California, in June 2004.

[Left] *The Americans used a Saturn IB rocket as launch vehicle to lift the Apollo portion of ASTP for rendezvous with the Russian Soyuz. This photograph, shot at dawn, shows the stack in place for a simulated launch sequence.*

[Below left] *Once the race to the Moon had been decided, the two superpowers—enjoying a period of détente during the 1970s—agreed to a cooperative venture in space that presaged the space station ideal of collaboration in orbit. The Apollo Soyuz Test Project (ASTP) represented a historic, if symbolic, meeting of the rival powers in the heavens. Two cosmonauts appear in this image, taken the morning of the Soviet ASTP launch at the Baikonur Cosmodrome in Kazakhstan, July 15, 1975: commander Aleksey Leonov (left), and Valeriy Kubasov.*

[Below] *On July 17, 1975, before the actual docking, one of the astronauts aboard the orbiting Apollo capsule captured this view of the Soyuz ship from above. The Soyuz spacecraft consisted of three main parts: the round orbital module (top), the bell-shaped descent vehicle (center), and the cylindrical instrument assembly module (from which the solar panels protrude).*

[Right] *The two vehicles mated on July 17, 1975. Command module pilot Vance Brand (accompanied by Thomas Stafford and Donald Slayton) peers from the hatchway leading from the Apollo command module into the docking module.*

[Far right] *Artist Paul Fjeld created this heroic painting of the Apollo and Soyuz spacecraft. The Apollo Command/Service module appears on the left, the Docking Module is in the center, and Soyuz on the right. Soviet and American design teams devised a system to interface the Docking Module with the existing Soviet docking system.*

[Below right] *During July 17 and 18, 1975, the cosmonauts and astronauts aboard the linked Soyuz and Apollo spacecraft visited one another four times. In this encounter, Soviet ASTP engineer Valeriy Kubasov (left) and Apollo commander Thomas Stafford collaborated on a photograph in the Soyuz Orbital Module.*

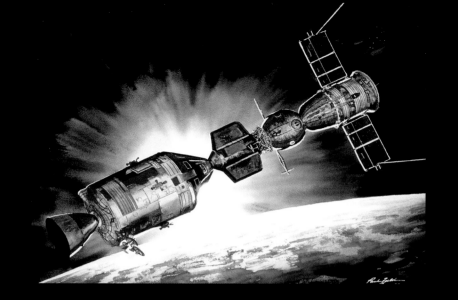

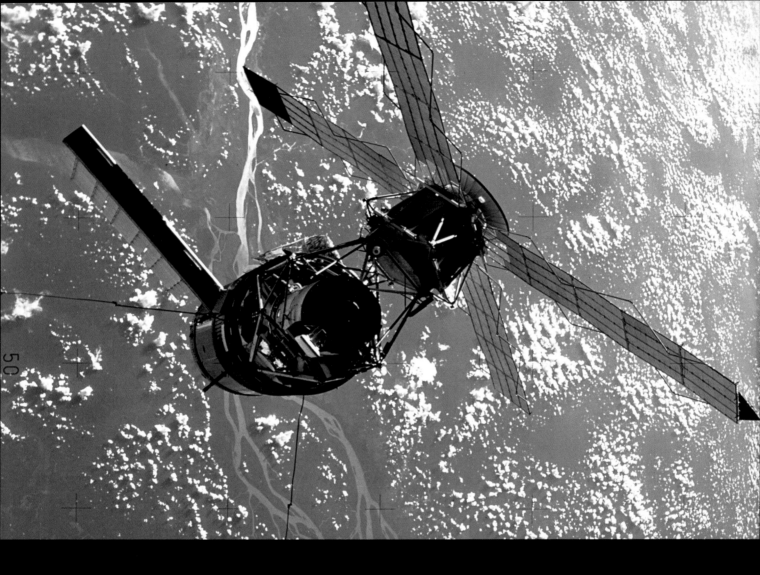

50

[Above] *If ASTP represented the international spirit associated with space stations,* Skylab *represented a separate strand in the concept, namely, long-endurance flight. Constructed from the upper stage of a Saturn V rocket, it mated with the Apollo Command/Service spacecraft.* Skylab *constituted an imaginative and frugal use of existing components. This photograph depicts the* Skylab *from the Apollo spacecraft, just prior to docking.*

[Opposite] Skylab 3 *astronauts in their spacecraft. The crew included (bottom left, picture left to right) commander Alan Bean, pilot Jack Lousma, and astronaut–scientist Owen Garriott. The three men spent fifty-nine days, eleven hours in orbit (late July to late September, 1974). Owen Garriott sits at one of the most important pieces of* Skylab *equipment, the Apollo Telescope Mount, a solar physics telescope. After conducting the medical and scientific experiments demanded by the* Skylab *mission, Jack Lousma takes a hot bath.*

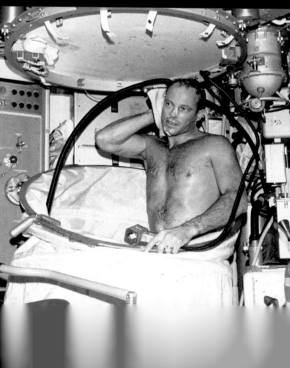

The Reagan administration (principally NASA Administrator James Beggs) backed an American space station like the one illustrated here. Marshall Space Flight Center assumed the main responsibility for the project after a Space Station Task Force rendered its findings during the early 1980s.

The Clinton administration and NASA Administrator Dan Goldin scaled back the dimensions of the space station, yielding a design shown in this artist's rendering of a new spacecraft called Alpha. During the 1990s, the role of the space station narrowed; no longer a symbol of Cold War prestige, it became little more than a laboratory and an astronomical observatory.

counterparts to restructure their independent space station plans and merge their efforts with NASA. Although this shared initiative failed to restore the space station to anything like its former size and multiple mission structure, the Russian contribution halted the long decline, and even added size to the structure. In fact, the two countries redesigned the *Alpha* station. In this new partnership, the foreign participation ceased to be merely important and instead became indispensable, woven into the architecture and planning itself. The Russians not only contributed new modules to the station but also increased the dimensions of the crew quarters and added length to the central corridor. Furthermore, the station's tarnished image brightened considerably with the increased capacity for passengers and greater electrical power generation. Moreover, the presence of the Russians increased the overall allure of the project, bringing much-needed drama to the undertaking and lending an appealing tone of goodwill between the once bitter rivals. Russian engineers and scientists also represented an important infusion of talent due to their impressive portfolios in designing and maintaining both the *Salyut* and the *Mir* vehicles. The third christening of the spacecraft—now the International Space Station (ISS)—reflected its new prestige.

This alliance between former enemies did not pay off immediately. Many in the American House of Representatives questioned the value of the Russian affiliation and in June 1993 the legislators defeated by just one vote (216–215) an amendment to end support for the orbiting laboratory. But by 1995 the beneficial effects of the partnership became undeniable. In that year, Congress approved a multi-year appropriation at the rate of a not insignificant $2,100,000,000 per year.

Russian and American space authorities adopted a conservative, incremental approach to their first major collaboration, calculated to integrate their contrasting styles of space operations at a measured pace. The first phase enabled American astronauts to experience long-duration life aboard an orbiting spacecraft. Seven U.S. crew members flew aboard the ageing *Mir* station between 1995 and 1998, enabling them to discover for themselves the effects of weightlessness for up to six months at a time. Meanwhile, seven cosmonauts flew on seven Shuttle missions during the same period.

Yet, the International Space Station remained a dyspeptic subject for many in Congress. Two emergencies aboard *Mir*, including an on-board fire, rang alarms in Washington, D.C. Far more aggravating, members learned in 1998 that starting in the previous year, NASA exceeded significantly the annual $2,100,000,000 space station budget limitation. An independent review predicted that ISS costs might surpass the 1993 estimate (of $17,400,000,000) by some $7,300,000,000. Additionally, the report projected development and deployment delays of from ten to thirty-eight months. Indeed, the year 2006, rather than the original 2004, now appeared to be the probable date of completion. Some noted with disbelief that a total of twenty-two years had passed since President Reagan had initiated plans for a permanent human presence in space.

Still, in spite of all of these obstacles, the ISS survived. Beginning in late 1998, the first American segment (the *Unity* module) mated with the initial Russian component (*Zarya*, or Sunrise). After long and politically embarrassing postponements, the second Russian section (*Zveda*, or

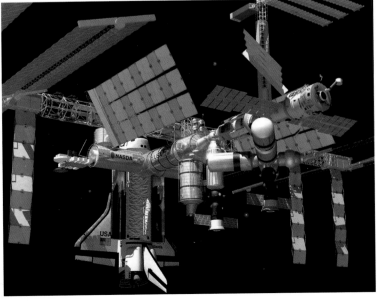

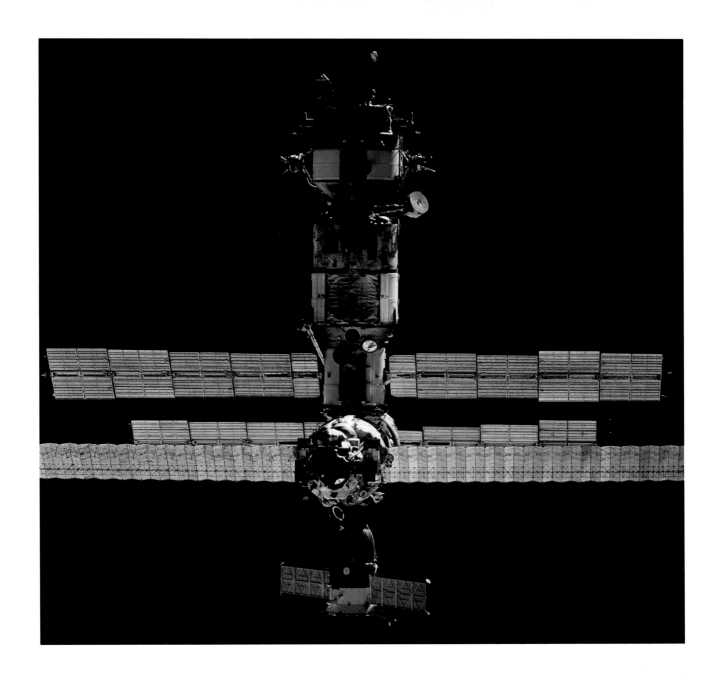

↑ *Daniel Goldin infused the concept of a space station with new enthusiasm, the prospect of a moderate enlargement in size, and the mantle of international cooperation by making the Russians full partners in the project. The* Mir *space station became the backbone of the new partnership. The Russians gained extensive experience maintaining and operating* Mir *during its many long-endurance missions, experience applicable to the International Space Station. Indeed, the collaboration began with dockings by* Mir *and the Space Shuttle in order to accustom the two sides to each other's practices—like the one in early 1995 between* Mir *(shown here) and the Shuttle* Discovery.

[Overleaf] *The Russians and Americans linked again in space in November 1995 when* Atlantis *(STS-74) and* Mir *formed an ad hoc international space station. After the two had docked, the astronauts looked out of the overhead windows of the orbiter to see the Russian spacecraft looming above.*

Star) joined with *Unity* in 2000, resulting in an ISS approximately 147 feet (45 m) in length and weighing 67 tons (68 tonnes). In November of that year, the first crews inhabited the station, taking four month rotations even as the construction of the ship proceeded around them. During the last phase of assembly, the European and Japanese partners planned to add their modules, bringing this immense machine to its full complement. The era that followed promised such practical (if uninspiring) pursuits as purifying pharmaceuticals in microgravity and determining prolonged exposure to weightlessness on human physiology. Bolder steps hinged, as they have since the dawn of spaceflight, less on technology and more on the political will of the nations involved.

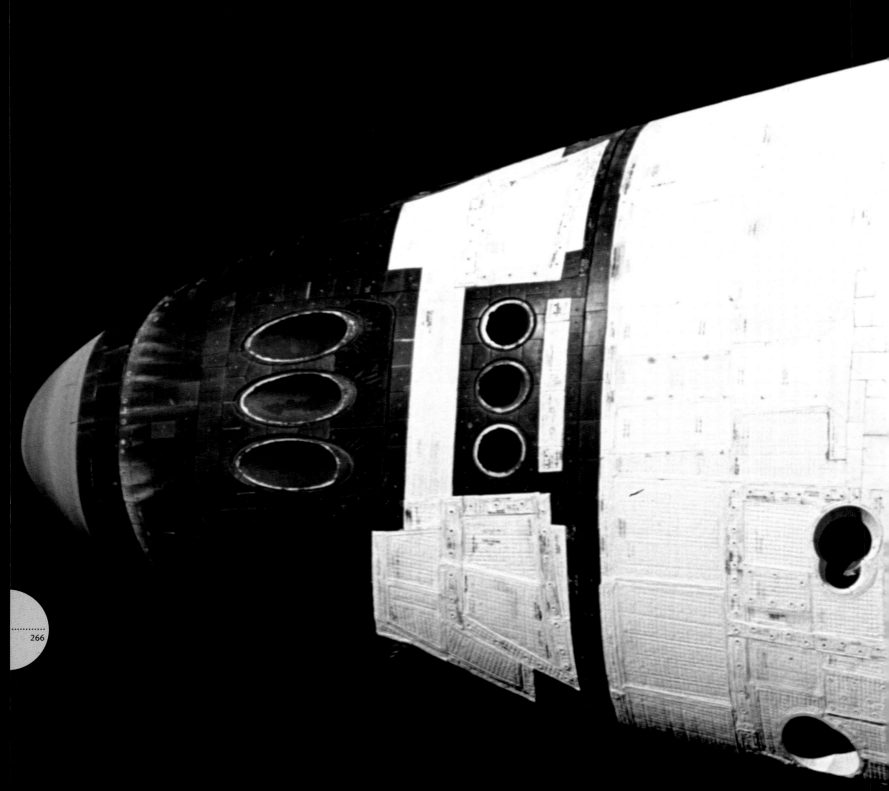

↑ *To achieve this powerful picture, the* Mir 19 *crew (Anatoliy Solovyev and Nikolai Budarin) disconnected the Soyuz spacecraft from the* Mir *and took a photograph of the unified* Mir–Atlantis. *Afterwards, the* Mir 18 *cosmonauts returned to Earth with* Atlantis.

At the beginning of the twenty-first century, the American space program found itself at a crossroads. Human travel outside the atmosphere became confined to Earth orbit, as the International Space Station and the Space Shuttle missions circled the planet again and again. The more provocative achievements fell to the drones. The Hubble Telescope captured images from the deepest recesses of the cosmos, and in early 2004 the Jet Propulsion Laboratory's Mars Rovers *Spirit* and *Opportunity* wheeled around on opposite sides of the Red Planet taking pictures, scooping soil samples, and promising unparalleled insights into the origins of the Earth's closest kin in the solar system.

Between the Hubble triumphs and the astounding success of the Rovers, human spaceflight suffered the unthinkable. The venerable Shuttle *Columbia*—the oldest orbiter in space and the one favored for science missions due to its capacity for longer flights—had been scheduled in 1998 to conduct microgravity and life sciences experiments in preparation for research aboard the International Space Station. Mission planners added a joint U.S.–Israeli space research project known as MEIDEX (Mediterranean–Israeli Dust Experiment), for which NASA selected an Israeli astronaut. Several factors delayed the mission, in the end a full two years past its January 2001 launch date. Other flights took priority (such as a service

←···· *The International Space Station began to take shape with the completion of the six sided central connector—known as* Unity—*that linked all subsequent pieces. Crew members of STS-88 examine the Boeing-made* Unity *(at this stage known as* Node 1*) in late 1997 at Kennedy's Space Station Processing Facility.*

F

[This page] *The first crew to inhabit the International Space Station stayed on board 138 days. Left to right, flight engineer Sergei Krikalev, commander William Shepard, and Soyuz commander Yuri Gidzenko, all wearing the Russian Sokol space suits. With their arrival in November 2000, the station underwent yet another change: the Shuttle* Endeavor *crew attached the 240 foot (73 m) long, 38 foot (11.6 m) wide solar array.*

[Opposite] *The fully assembled International Space Station (as of March 2001), viewed from the aft deck of the Shuttle* Discovery. *A cosmonaut stands by a window of the module* Zvezda. *An astronaut (flight engineer Susan Helms) mounts a video camera in the module* Zarya.

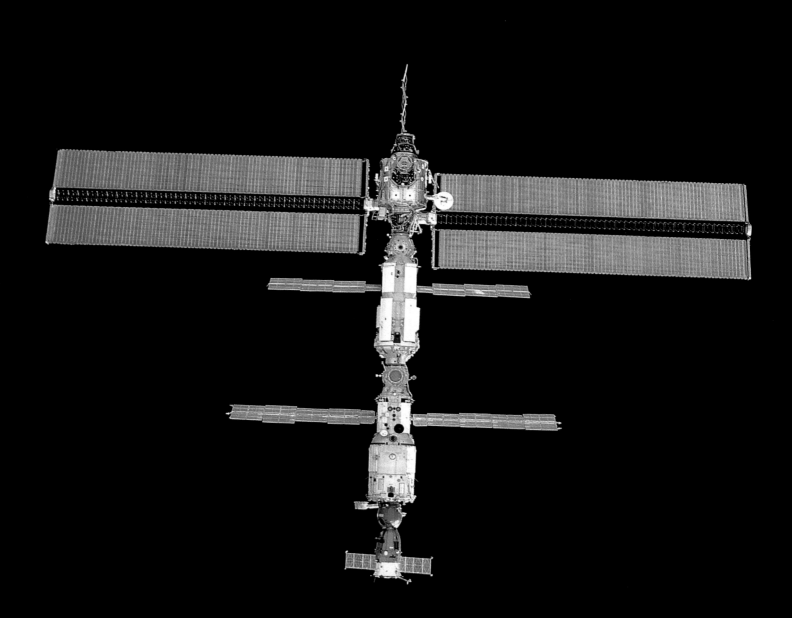

↑ *During early December 1998, the Shuttle* Endeavor *astronauts released* Unity *(right) from the cargo bay and connected it to the Russian space station component* Zarya, *launched earlier on a Russian Proton rocket.* Zarya *comprised the space station's Functional Cargo Block.*

visit to the Hubble Telescope) and *Columbia* underwent fundamental maintenance that took six months longer than expected.

Meanwhile, the flight selected to carry out these assignments—STS-107—fielded a crew trained to conduct a densely packed cluster of experiments. No different from other shuttle missions, this one carried seven relatively young men and women of distinguished accomplishments. Its commander, Colonel Rick Husband, had been an Air Force test pilot and had flown on STS-96. Just forty-five years of age, he had received bachelor's and master's degrees in mechanical engineering. *Columbia*'s pilot, forty-one-year-old Navy commander William C. McCool, another test pilot, possessed a B.S. in applied science, as well as twin master's degrees in computer

science and aeronautical engineering. The crew had two medical doctors. Mission specialist Laurel Clark, also forty-one years old, had recently been selected to be a captain in the U.S. Navy where she had been a flight surgeon. Captain David Brown, another mission specialist, had served as a naval flight surgeon, as well as an aviator.

The STS-107 team included two members with advanced backgrounds in the physical sciences and engineering. Michael Anderson, payload commander and mission specialist, pursued a career in the U.S. Air Force as a lieutenant colonel, had been a flight instructor, and had flown on STS-89. Anderson, aged forty-three, earned a bachelor's degree in physics and astronomy and a master's degree in physics. Flight engineer and mission specialist Kalpana Chawla, forty-one, held bachelor's,

↑ *The third portion of the International Space Station materialized in September 2000 when the astronauts of STS-106 linked a second Russian component,* Zvezda *(a service module), to* Zarya. *The picture of the three-part space station also includes the Russian supply vehicle* Progress, *at the far end of the space station.*

master's, and doctoral degrees in aerospace engineering, had flown aboard STS-87, and had received a flight instructor's certification from the Federal Aviation Administration. Finally, forty-eight year old payload specialist Ilan Ramon, an Israeli Air Force colonel and fighter pilot, became his country's first astronaut upon appointment to STS-107.

Nothing out of the ordinary happened as the crew of STS-107 awaited the launch of *Columbia* during the early morning of January 16, 2003. The massive shuttle stack had been drenched in a cloudburst as it stood on Pad 39A at the Kennedy Space Center, but suffered no ill effects. Frost had been detected the morning of the launch, but it had dissipated by 7:15 am. Fifteen minutes later the seven astronauts embarked for *Columbia*, and within two hours

they had been sealed in for the long ride. At 10:39 am the solid rocket boosters ignited and the machine began to rise. At fifty-seven seconds into the voyage the spacecraft underwent buffeting due to windshear, but the loads on the various structural members failed to exceed design limits. One hour after launch *Columbia* began circling the Earth and during the sixteen days aloft, the crew undertook many beneficial experiments. Ilan Ramon turned the MEIDEX equipment on thunderstorms over central Africa and later tracked an immense dust storm over the Atlantic Ocean for three days. Others tested bone cells, prostate cancer, and bacterial growth in weightlessness. The astronauts' bodies underwent screenings for protein production, bone and calcium manufacture, and renal stone formation in microgravity.

→ The crew of STS-107 in front of a T-38 jet trainer at Ellington Field near Johnson Space Center. The astronauts are (from left) commander Rick Husband, pilot William McCool, mission specialists David Brown and Laurel Clark, payload specialist Ilan Ramon, and mission specialists Michael Anderson and Kalpana Chawla.

↓ ↘ → Two Columbia astronauts (Laurel Clark pointing, Kalpana Chawla beside her, right) receive training on some of the science experiments to be placed aboard STS-107. Israeli payload specialist Ilan Ramon conducts a SPACEHAB experiment. After final preparations, STS-107 lifts off from Pad 39A on January 16, 2003.

Astroculture experiments entailed the harvesting of oil samples from roses and rice flowers. Finally, on Day 17 (Saturday, February 1) the crew shut the payload doors and the commander and pilot activated the re-entry software in *Columbia*'s computer.

At a little after 8:15 am Eastern Standard Time on Saturday, Rick Husband began the de-orbit burn 175 miles (282 km) over the Indian Ocean. For two and a half minutes the orbiter's twin orbital maneuvering system engines slowed the spacecraft as it flew upside down and tail first. Once the maneuver was completed, Husband righted the ship and flew it toward the Earth in a nose-up attitude. Half an hour later *Columbia* entered the atmosphere at about 400,000 feet (122,000 m) and the astronauts saw bright flashes engulfing the vehicle, normal as leading-edge temperatures approached 2500° F (1371° C). Less than five minutes later, a sensor on the left side of the leading-edge spar indicated higher than normal strain.

Then casual observers on the ground saw disquieting omens in the dark western sky. Crossing the California coastline, as wing leading edges heated to about 2800° F (1538° C), the orbiter's light track appeared to brighten and shed debris at about 8:54 am (5:54 am Pacific Time). Over the next twenty-three seconds witnesses reported four similar events, and then a flash as the spacecraft entered Nevada airspace flying at Mach 22.5 and at a height of 227,400 miles (365,900 km). More of the same continued as *Columbia* flew over Utah, Arizona, and New Mexico toward Texas. At Mission Control in Houston, the first ominous signs manifested themselves. At a little past 8:54 the engineers learned that four sensors in the left wing had failed. Meanwhile, a thermal protection tile fell to the ground near the border of New Mexico and Texas, the first known place where debris landed.

Five minutes after the wing sensors quit, pressure readings flattened on both of the tires on the left main landing gear. Seventeen seconds later, the commander's voice crackled through the communication system for the last time. Just after the scheduled landing time of 9:16 am Eastern Time passed, the agency declared an emergency. At 10:30 that morning, NASA Administrator Sean O'Keefe, who had been at Kennedy Space Center awaiting the return of *Columbia* with the families of the crew members, announced the appointment of an accident investigation board chaired by retired Admiral Harold W. Gehman, Jr.

Mr. O'Keefe named Admiral Gehman just after activating an interim federal entity known as the International Space Station and Space Shuttle Mishap Interagency Investigation Board, by charter consisting of seven members based on their positions in government. The panel comprised: Major General Kenneth W. Hess, Chief of Safety, U.S. Air Force; Mr. Steven B. Wallace, Director, Office of Accident Investigation, Federal Aviation Administration; Brigadier General Duane W. Deal, U.S. Air

Harold W. Gehman, Jr.

Admiral Harold Webster Gehman, Jr., entered the U.S. Navy in the Reserve Officers' Training Candidates program at Pennsylvania State University, where he earned a B.S. degree in industrial engineering. Born in Norfolk, Virginia, to a navy captain in October 1942, Gehman served for thirty-five years in five sea commands, rising in rank from lieutenant to rear admiral. He began with two tours in Vietnam, where he commanded a Swift patrol boat and later a detachment of six Swifts. He subsequently commanded the guided missile destroyer *Dahlgren* and the guided missile cruiser *Belknap*. During the administrative part of his career, Gehman served four times on the staff of the Chief of Naval Operations, and became deputy of the Navy's Atlantic Command. He earned his fourth star in 1996, becoming the 29th Vice Chief of Naval Operations. At the time of retirement, Admiral Gehman led the U.S. Joint Forces Command. Thoughtful, deliberate, and self-possessed, Gehman had previous experience chairing high profile investigations. In 2000 President Clinton asked him to lead a commission to determine the causes of the attack on the USS *Cole*, a destroyer that had been refueling in Aden harbor in Yemen when terrorists detonated a massive explosion beside her, tearing a 40 by 40 foot (12 by 12 m) cavern in her hull. This assignment had burnished Gehman's reputation as a fact-finder, but in addition, NASA Administrator O'Keefe, Secretary of the Navy between 1992 and 1993, knew and respected Gehman from earlier associations, rendering him a logical choice to guide the complex and emotionally charged *Columbia* inquiry.

Force Space Command; Rear Admiral Stephen A. Turcotte, Commander, U.S. Navy Safety Center; Dr. James N. Hallock, Director, Aviation Safety Division, Department of Transportation; Major General John L. Barry, U.S. Air Force Materiel Command; and G. Scott Hubbard, Director, NASA Ames Research Center.

During February and March, Gehman expanded the panel, adding representatives from the nation's universities. The admiral appointed Dr. Douglas D. Osheroff, Professor of Physics and Applied Physics at Stanford University (1996 Nobel Laureate in Physics); Dr. Sheila E. Widnall, Professor of Aeronautics and Astronautics at MIT (and former Secretary of the US Air Force); Dr. Sally T. Ride, Professor of Physics, University of California at San Diego (the first American woman astronaut, and a member of the Challenger Accident Investigation Board); Dr. John M. Logsdon, Director of the Space Policy Institute, George Washington University; and Mr. Roger E. Tetrault, retired Chairman and CEO of McDermott International.

The Columbia Accident Investigation Board (or CAIB, as it came to be known), which supplanted the Mishap Interagency Investigation Board, began its task by sifting and analysing a mountain of evidence, much of it the fallen remains from the orbiter itself. A 2000 square mile (5180 square km) debris field, shaped like a cigar, ranged from Fort Worth, Texas, to Fort Polk, Louisiana, deposited as *Columbia* disintegrated. At first, National Guard, Texas Department of Public Safety, and local emergency responders conducted ground searches. Soon, the Forest Service became engaged, supplying most of the equipment and personnel, particularly wild land firefighters. In all, about 25,000 individuals participated, recovering some 84,000 pieces of *Columbia* and combing on foot a 700,000 acre (280,000 ha) landscape.

⟨⋯⟩ *On February 1, 2003, NASA Administrator Sean O'Keefe (left) appointed a board to investigate the* Columbia *tragedy, and did so a little more than an hour after the orbiter and crew failed to return to Kennedy Space Center. He named Harold W. Gehman (right), a retired admiral, to chair the committee. Gehman started work that day.*

Service mapped out and then walked a grid during the search for Columbia debris near Hemphill, Texas. One NASA official accompanied each of these teams in order to iden hazardous materials, if need A massive outpouring of priv volunteers, as well as those state, local, and federal governments, helped comb t enormous zone in which part from the sky. Some pieces to little searching. An 800 poun (363 kg) unit from one of the Shuttle main engines droppe into Fort Polk, Louisiana.

Several hangars came to be
reconstruction sites for wreckage
from the Columbia accident. At
Kennedy Space Center, the floor
of the Reusable Launch Vehicle
hangar served as a grid on which
to lay out recovered parts. In
particular, the Columbia Board
wanted to reconstruct the bottom
of the orbiter. Another hangar, at
Barksdale Air Force Base in
Shreveport, Louisiana, acted
as a holding point for collected
materials before being sent
to Kennedy.

Investigators learned much from the cache. They found the modular auxiliary data system, a recorder containing data not transmitted by telemetry to Mission Control, including temperature sensor readings from the left wing leading edge. They also found it possible to assemble a three-dimensional model of the affected left wing by analysing the damage evident on the surviving pieces and studying the order in which they had fallen from the sky. Based on this evidence, the investigators developed a persuasive theory about the physical origin of the accident.

Thus, at a relatively early point in their inquiry, the CAIB panelists thought that the *Columbia* catastrophe had resulted from a failure of the left wing. But what had started it? Some blamed the insulating foam covering the Shuttle's external fuel tank. But time and again in previous Shuttle flights, foam had struck parts of the stack, always without incident. Indeed, before the disaster, engineers at Johnson Space Center had dismissed the prospect of foam causing a breach in the

orbiter due to these past benign experiences. In fact, foam loss had occurred on sixty-three of the seventy-nine shuttle missions for which imaging provided evidence.

But members of the CAIB reconsidered this received wisdom. Combining the physical evidence with the immense amount of data collected by Houston during the fateful end of *Columbia*'s mission, scientists on and off the board pondered whether a material as light and soft as insulation might have brought down the biggest spacecraft ever to fly. They answered the question using a variety of techniques. Computational fluid dynamics approximated the trajectory of the foam when it struck the orbiter. Image analysis offered clues about the size, speed, origin, and strike area of the piece in question. Finally, a 300 foot (91 m), nitrogen-powered gun located at the Southwest Research Institute fired foam test articles at sections of the orbiter's wing, simulating the speed and flight environment as it occurred during the launch of STS-107. Based on this research, members of the Columbia Accident Investigation Board were able to state

THE PHYSICAL CAUSE OF THE LOSS OF *COLUMBIA* AND ITS CREW WAS A BREACH IN THE THERMAL PROTECTION SYSTEM ON THE LEADING EDGE OF THE LEFT WING.

←--- ↓ The loss of Columbia *exacted an emotional toll on the nation. An impromptu memorial sprang up around the entrance sign at the Johnson Space Center. On February 4, 2004, President George W. Bush spoke eloquently to Johnson employees, the families of the deceased, and the nation, about the high price that exploration sometimes demands. A little less than a year later, the president declared a new age of space exploration in which the Moon and Mars once again figured prominently in American objectives in space.*

confidently the cause of the *Columbia* accident in their August 2003 report:

The physical cause of the loss of Columbia *and its crew was a breach in the Thermal Protection System on the leading edge of the left wing. The breach was initiated by a piece of insulating foam that separated from the left bipod ramp of the External Tank and struck the wing in the vicinity of the lower half of Reinforced Carbon-Carbon panel 8 at 81.9 seconds after launch. During re-entry, this breach in the Thermal Protection System allowed superheated air to penetrate the leading-edge insulations and progressively melt the aluminum structure of the left wing, resulting in a weakening of the structure until increasing aerodynamic forces caused loss of control, failure of the wing, and breakup of the Orbiter.[3]*

But Admiral Gehman and his colleagues did not content themselves with uncovering the technical roots of the *Columbia* loss. They also cited flaws in the manner in which the U.S. space agency conducted itself as an institution, which reflected more systemic problems. The group discussed the compromises in Shuttle design that had been necessary for it to win congressional and presidential approval, and the impact these had had on Shuttle safety. They pointed out the ill-effects of prolonged funding constraints, schedule pressures, and inconsistent priorities, all of which had contributed to increased risk. As the Shuttle became a mature technology with a long history, many in the program began to substitute past success for critical thinking, and there was evidence that some individuals suppressed professional differences of opinion when the disagreement conflicted with programmatic imperatives. Moreover, some of the people who attempted to raise safety concerns found themselves at odds with institutional dogma.

Finally, Gehman and his team leveled two broad complaints, one targeted at NASA, the other more generalized. They argued that the agency had deceived itself into believing that the Space Shuttle resembled an operational spacecraft—like an airliner—when the facts of its launch and maintenance showed that even after its hundredth flight it remained a developmental vehicle, much like the X-15 of an earlier generation. The original metaphor of a space truck persisted, even though the Shuttle behaved more like a concept car. Additionally, the board observed that the American space program lacked a broad set of national objectives to inform its year by year, programmatic decision-making.

Even before NASA announced a date for the Space Shuttle's return to flight, President George W. Bush—no doubt encouraged by the stunningly successful landing of the Mars Rover *Spirit* on January 4, 2004 (*Opportunity* followed later that month)—announced on January 14, 2004, a new path for the American space program. In words reminiscent of those of President John F. Kennedy spoken more than forty years earlier, Mr. Bush said, "We will give NASA a new focus and vision for future exploration. We will build new ships to carry man forward into the universe, to gain a new foothold on the Moon, and to prepare for new journeys to worlds beyond our own."[4]

With this declaration, he inaugurated the development of a new spacecraft known as the Crew Exploration

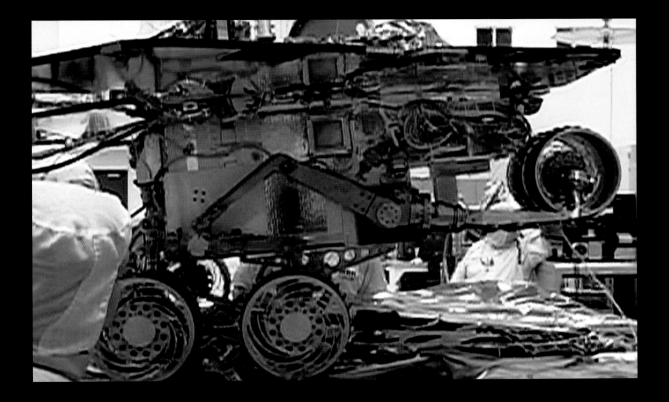

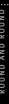

[Opposite] *The Rover* Opportunity *responds to ground controllers after landing. Here, facing opposite to the direction in which it finally moved,* Opportunity *prepares to roll free of the crumpled Lander, and extend its instrument deployment device (or arm).*

[Above] *President Bush announced his space initiative at least in part because of two space triumphs. In January 2004, Mars Rovers* Spirit *and* Opportunity *not only landed safely on the Red Planet, but roamed on command. On Earth, engineers at the Jet Propulsion Laboratory subjected the Rovers to countless tests, including the one shown here in which the stand-up motions underwent scrutiny.*

[Right] *The Rover* Spirit *transmitted pictures to JPL just after its landing. Shown here is a portion of the very first color picture captured by its panoramic camera.*

An incomplete panoramic picture shows Opportunity's Lander (right) rests inside a crater. Also seen are Rover tracks at the center of the image, and airbag bounce marks behind the Lander. In the artist's conception opposite, the Rover as it appears on the Red Planet.

Vehicle (CEV), proposed a series of robotic landings on the moon starting in 2008, declared the end of the Space Shuttle's operational life in 2010, challenged NASA to launch the CEV's first human mission in 2014, and to land on the Moon between 2015 and 2020. After establishing the Moon as a base camp, the president spoke hopefully of flying human beings from there to Mars, preceded by a host of robotic missions. Less than three weeks later, NASA headquarters reorganized itself to form the Office of Exploration Systems, dedicated to fulfilling the objectives outlined by the president.

Whether all these ambitious desires will come to fruition remains uncertain. Other presidents, including Mr. Bush's own father, have made sweeping declarations only to find themselves and Congress beset by the kinds of budgetary pressures that prohibited anything more than the piecemeal space program criticized by Admiral Gehman's board. Whether this initiative proves to be successful or not, as NASA celebrates its fiftieth anniversary in 2008 (and with the venerable NACA set to celebrate its hundredth birthday in 2015, the year Mr. Bush hopes to return Americans to the Moon), the US electorate will ultimately decide, through their political representatives, the scope and direction of future celestial voyages of discovery.

1. This project, later known as the Apollo Telescope Mount, flew aboard the *Skylab* space platform during the 1970s. It is discussed later in this chapter in the context of the International Space Station.

2. *Public Papers of the Presidents of the United States: Ronald Reagan, 1984*, quoted in Launius, *NASA: A History of the U.S. Civil Space Program*, p. 248.

3. CAIB Report, Volume I, p. 49.

4. The White House, Office of the Press Secretary, January 14, 2004: "President Bush Announces New Vision for Space Exploration Program."

1903
Wright brothers' first powered flight

1915
The National Advisory Committee for Aeronautics (NACA) comes into being as a part of the Naval Appropriations Act

1922
The Variable Density Wind Tunnel, conceived by German émigré scientist Max Munk, begins operation and wins acclaim for the NACA

1940
NACA's Ames Aeronautical Laboratory opens in California

1942
NACA's Aircraft Engine Laboratory dedicated in Ohio

1947
John Stack and the NACA (collaborating with the U.S. Air Force) design and flight test the supersonic Bell X-1

Hugh L. Dryden succeeds George W. Lewis as Director of the NACA

1926-1930
NACA pursues pressure distribution research

1919
George W. Lewis is appointed NACA's first director

| 1900 | 1920 | 1930 | 1940 |

1920
The NACA's Langley Memorial Aeronautical Laboratory opens in Hampton, Virginia

NACA dedicates its first wind tunnel

1934
Langley aerodynamicist John Stack publishes a landmark paper about supersonic flight in The Journal of the Aeronautical Sciences

1926-1929
Experiments by Langley researcher Fred E. Weick result in engine cowlings that enable significant reductions in drag without causing engine overheating

1948
NACA pilot Howard Lilly dies in a crash of a Douglas D-558 Skyrocket

1957
The U.S.S.R. wins the first round of the space competition, placing Sputnik I (shown here in the assembly shop) in orbit and causing an outcry in the U.S.

1958
In January the U.S. answers Sputnik with Explorer 1 (below), the first American satellite sent aloft, followed by Vanguard 1 in March

1953
Flying the Douglas D-558 Skyrocket, NACA pilot A. Scott Crossfield becomes the first human to exceed Mach 2

U.S. Air Force unveils Dyna-Soar, a boost-glide hypersonic aircraft

1950

1954
The X-15 program begins

The International Council of Scientific Unions initiates the International Geophysical Year, which occasions the space race between the U.S. and the U.S.S.R. Below, a U.S. Viking rocket undergoing static testing

1958
Project Mercury begins under Robert Gilruth, a NACA engineer who headed the Space Task Group at Langley

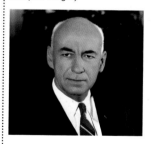

Out of the shell of the venerable NACA, the National Aeronautics and Space Administration emerges on October 1

1959
Scott Crossfield becomes the first to fly the X-15 (after being the first to exceed Mach 2 in 1953), but almost crashes on landing

Astronaut training begins

Timeline

1965

Mariner 4 *flies past Mars, taking twenty-one pictures of the planet*

During the flight of Gemini 4, astronaut Edward White (left) becomes the first American to undertake an extravehicular activity (EVA)

1962

Mariner 2 *swings past Venus and enters deep space*

Project Gemini authorized

Manned Spacecraft Center opens in Houston, Texas

John Glenn becomes the first American to orbit the Earth, but experiences an errant autopilot and faulty heat shield indicator

1961

Cosmonaut Yuri Gagarin orbits the Earth in Vostock I

President Kennedy announces Apollo on May 25 and pledges to send astronauts to the Moon and back within the decade

1968

The final X-15 mission takes place in October, ending flight research on the world's fastest piloted aircraft

In December, Apollo 8 circumnavigates the Moon

1966

Surveyor *spacecraft lands on the Moon in June*

1960

1961

Suborbital flights are made by astronauts Alan B. Shepard (shown below; flew on May 6) and Gus Grissom (July 21). Both flights are fraught with technical difficulties

1964

Ranger 7 *photographs the Moon before crashing into it*

1963

First flight of the M2-F1 lifting body, prototype of high lift-to-drag vehicles that enable runway landings from space

Assassination of President Kennedy, November 22

1967

Apollo 1 *astronauts perish when their capsule catches fire*

The Saturn V rocket—a 363 foot (111 m) behemoth capable of six million pounds of thrust—makes its maiden flight

1970
Apollo 13 *mission ends after near catastrophe*

1973
Pioneer 10 *(launched in March 1972) and* Pioneer 11 *(April 1973) leave the solar system after passing Jupiter, Saturn, and Neptune. Below, the craft in a space simulation chamber*

1969
Edwin (Buzz) Aldrin (below) and Neil Armstrong walk on the Moon during the Apollo 11 *mission as Michael Collins waits in lunar orbit*

1971
NASA and the U.S. Air Force begin their Space Shuttle collaboration

1970

1969
Report issued by President Nixon's Space Task Group envisions a sweeping period of space exploration. Below, President Nixon speaks to the astronauts on the Moon

1972
President Nixon gives the go-ahead for the Space Shuttle

The U.S. lunar program ends with Apollo 17

Mariner 9 (shown here being examined prior to encapsulation) completes its survey of 85% of the Martian surface. The results offer hope of finding evidence for past life on the planet

1974
The last Skylab mission ends in February. Like the first two Skylabs, it prepares NASA for long duration flight

289

1975
Vikings 1 and 2 set sail for Mars in August and September, respectively. Both land on the Red Planet, taking color photos. Below, Viking 2 undergoing assembly in a cleanroom

1977
Voyager 2, launched in August, embarks on a "Grand Tour" of the solar system, flying past Jupiter, Saturn, Uranus, and Neptune. Below, a technician installs a record containing typical sounds heard on Earth

1978
Go-ahead for the Hubble Space Telescope

1979
Orbiter Columbia leaves Palmdale, California, for the Kennedy Space Center

1983
Sally K. Ride becomes the first U.S. woman in space, flying aboard STS-7 as a mission specialist

1980

1981
The inaugural flight of the Shuttle occurs on April 12; Columbia circles the Earth 36 times before returning home

1977
Between August and October, the Shuttle orbiter Enterprise undergoes unpowered flight tests at the NASA Dryden Flight Research Center

1988
Discovery *returns the Shuttle program to flight*

1995
Congress commits itself to the International Space Station (ISS)

2004
President George W. Bush announces in January a space exploration initiative to return Americans to the Moon

Also in January, the two Mars Rovers landed on the Red Planet. Below, a Rover spacecraft's stand-up motions are tested

1990
Hubble Space Telescope (shown here undergoing assembly) goes into operation in April, only to be plagued by blurred vision

2000
Russian component Zveda *joined to the ISS*

2001
Orbiter Columbia *scheduled for an important science mission*

1985
NASA unveils its initial space station plan

Mars Rovers Spirit *and* Opportunity *roam on opposite sides of the planet*

1990

2000

1993
Hubble servicing mission restores the telescope's sight

2003
After two years' delay, the science mission aboard Columbia *lifts off. On February 1,* Columbia *disintegrates over the southwest U.S.*

1998
Initial mating of the ISS components, linking the U.S. Unity *and the Russian* Zarya *segments*

1986
Shuttle orbiter Challenger *explodes, disintegrates, and falls into the Atlantic Ocean about a minute after launch on January 28*

The Rogers Commission presents its findings on Challenger *some four months after the disaster*

Release of the Columbia Accident Investigation Board report (August)

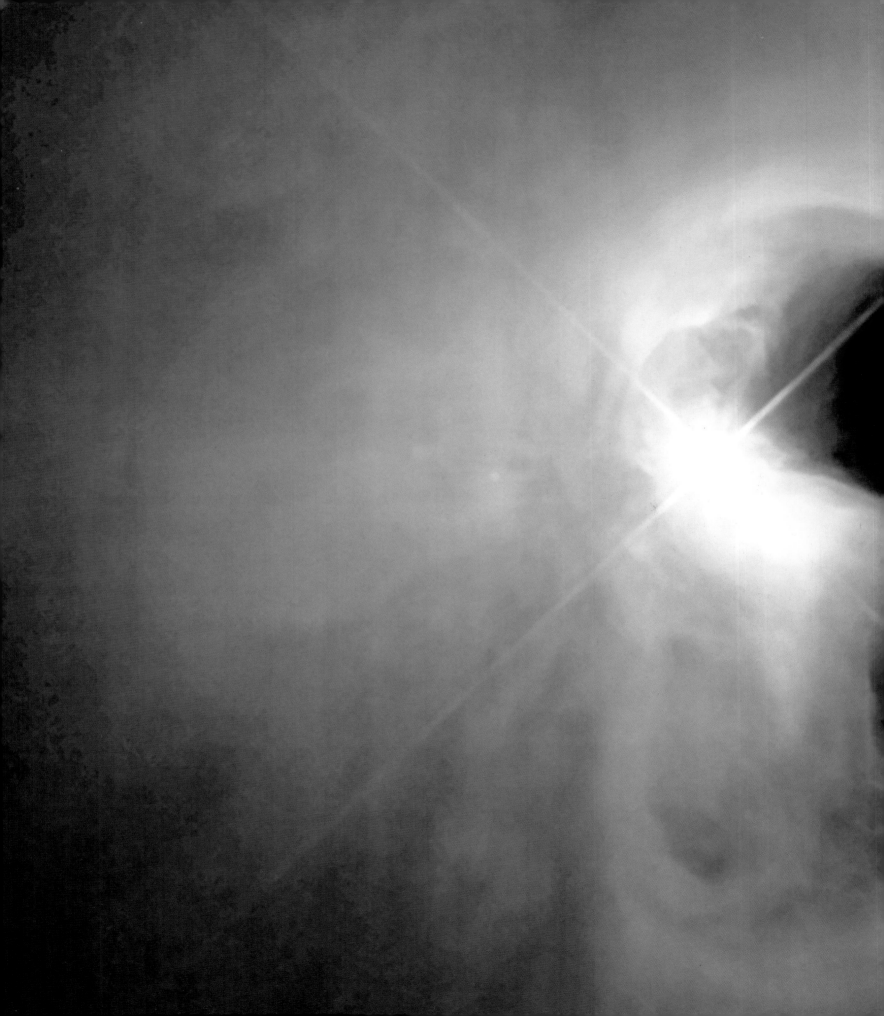

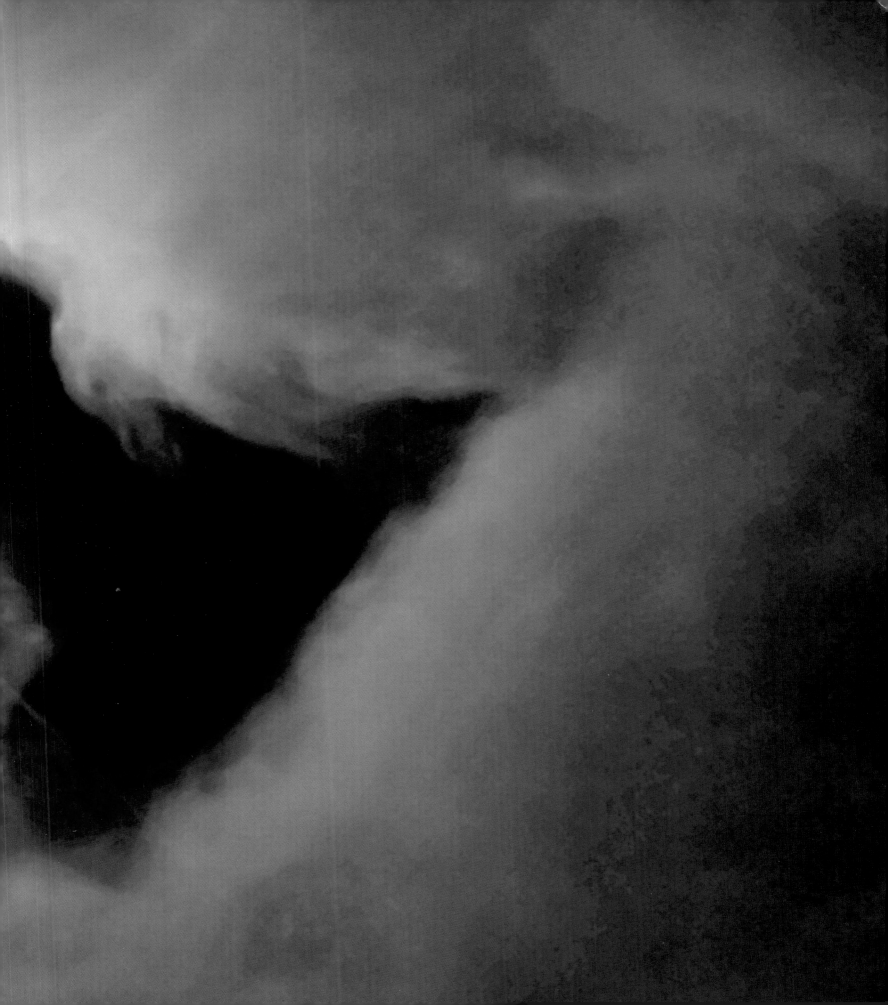

Aldrin, Edwin E., *Return to Earth*, New York (Random House) 1973

Beattie, Donald A., *Taking Science to the Moon: Lunar Experiments and the Apollo Program*, Baltimore and London (The Johns Hopkins University Press) 2001

Becker, John V., *The High-Speed Frontier: Case Histories of Four NASA Programs, 1920–1950*, NASA Special Publication 445, Washington, D.C. (NASA) 1980

Bilstein, Roger E., *Stages to Saturn: A Technological History of the Apollo/Saturn Launch Vehicles*, NASA Special Publication 4206, Washington, D.C. (NASA) 1996

Bilstein, Roger E., *Testing Aircraft, Exploring Space: An Illustrated History of NACA and NASA*, Baltimore and London (The Johns Hopkins University Press) 2003

Bromberg, Joan Lisa, *NASA and the Space Industry*, Baltimore and London (The Johns Hopkins University Press) 1999

Bugos, Glenn E., *Atmosphere of Freedom: Sixty Years at the NASA Ames Research Center*, NASA Special Publication 4314, Washington, D.C. (NASA) 2000

Burrows, William E., T*he Infinite Journey: Eyewitness Accounts of NASA and the Age of Space*, New York (Discovery Books) 2000

Butrica, Andrew J., ed., *Beyond the Ionosphere: Fifty Years of Satellite Communication*, NASA Special Publication 4217, Washington, D.C. (NASA) 1997

Collins, Martin J., and Sylvia K. Kraemer, eds., *Space: Discovery and Exploration*, New York (Hugh Lauter Levin) 1994

Collins, Michael, *Liftoff: The Story of America's Adventure in Space*, New York (Grove Press) 1988

Compton, William David, *Where No Man Has Gone Before: A History of Apollo Lunar Exploration Missions*, NASA Special Publication 4214, Washington, D.C. (NASA) 1989

Cortright, Edgar M., ed., *Apollo Expeditions to the Moon*, NASA Special Publication 350, Washington, D.C. (NASA) 1975

Dawson, Virginia P., *Engines and Innovation: Lewis Laboratory and American Propulsion Technology*, NASA Special Publication 4306, Washington, D.C. (NASA) 1991

Dethloff, Henry C., *Suddenly Tomorrow Came: A History of the Johnson Space Center*, NASA Special Publication 4307, Washington, D.C. (NASA) 1993

Dethloff, Henry C., and Ronald A. Schorn, *Voyager's Grand Tour: To the Outer Planets and Beyond*, Washington, D.C. and London (Smithsonian Institution Press) 2003

Dunar, Andrew J., and Stephen P. Waring, *Power to Explore: A History of Marshall Space Flight Center, 1960–1990*, NASA Special Publication 4313, Washington, D.C. (NASA), 1999

Glenn, John, *A Memoir*, New York (Bantam Books) 1999

Glennan, T. Keith, *The Birth of NASA: The Diary of T. Keith Glennan*, NASA Special Publication 4105, Washington, D.C. (NASA) 1993

Gorn, Michael H., *Expanding the Envelope: Flight Research at NACA and NASA*, Lexington, Ky. (The University Press of Kentucky) 2001

Gorn, Michael H., *Hugh L. Dryden's Career in Aviation and Space*, NASA Monographs in Aerospace History Number 5, Washington, D.C. (NASA), 1996

Greenwood, John T., ed., *Milestones of Aviation*, New York (Hugh Lauter Levin) 1989

Hacker, Barton C., and James M. Grimwood, *On the Shoulders of Titans: A History of Project Gemini*, NASA Special Publication 4203, Washington, D.C. (NASA) 1977

Hallion, Richard P., and Michael H. Gorn, *On the Frontier: Experimental Flight at NASA Dryden*, Washington, D.C. and London (Smithsonian Books) 2003

Hansen, James R., *Enchanted Rendezvous: John C. Houbolt and the Genesis of the Lunar-Orbit Rendezvous Concept*, NASA Monographs in Aerospace History Number 4, Washington, D.C. (NASA) 1995

Hansen, James R., *Engineer in Charge: A History of the Langley Aeronautical Laboratory, 1917–1958*, NASA Special Publication 4305, Washington, D.C. (NASA) 1987

Hansen, James R., *Spaceflight Revolution: NASA Langley Research Center from Sputnik to Apollo*, NASA Special Publication 4308, Washington, D.C. (NASA) 1995

Heppenheimer, T.A., *Development of the Shuttle, 1972–1981*, Washington, D.C. and London (Smithsonian Institution Press) 2002

Heppenheimer, T.A., *The Space Shuttle Decision: NASA's Search for a Reusable Launch Vehicle*, NASA Special Publication 4221, Washington, D.C. (NASA) 1999

Herring, Mack R., *Way Station to Space: A History of the John C. Stennis Space Center*, NASA Special Publication 4310, Washington, D.C. (NASA) 1997

Huffbauer, Karl, *Exploring the Sun: Solar Science Since Galileo*, Baltimore and London (The Johns Hopkins University Press) 1991

Jenkins, Dennis R., *Hypersonic: The Story of the North America X-15, North Branch*, Minn. (Specialty Press) 2003

Jenkins, Dennis R., *Hypersonics Before the Shuttle: A Concise History of the X-15 Research Airplane*, NASA Monographs in Aerospace History Number 18, Washington, D.C. (NASA) 2000

Jenkins, Dennis R., *Space Shuttle: A History of the National Space Transportation System*, Stillwater, Minn. (Voyageur Press) 2002

Johnson, Stephen B., *The Secret of Apollo: Systems Management in American and European Space Programs*, Baltimore and London (The Johns Hopkins University Press) 2002

Kelly, Thomas J., *Moon Lander: How We Developed the Apollo Lunar Module*, Washington, D.C. and London (Smithsonian Institution Press) 2001

Kraemer, Robert S., *Beyond the Moon: A Golden Age of Planetary Exploration, 1971–1978*, Washington, D.C. and London (Smithsonian Institution Press) 2000

Lambright, W. Henry, *Powering Apollo: James E. Webb of NASA*, Baltimore and London (The Johns Hopkins University Press) 1995

Lambright, W. Henry, *Space Policy in the Twenty-First Century*, Baltimore and London (The Johns Hopkins University Press) 2003

Launius, Roger D., *NASA: A History of the U.S. Civil Space Program*, Malabar, Fla. (Krieger Publishing) 1994

Launius, Roger D., and Dennis R. Jenkins, eds., *To Reach the High Frontier: A History of U.S. Launch Vehicles*, Lexington, Ky. (The University Press of Kentucky) 2002

Launius, Roger D., John M. Logsdon, and Robert W. Smith, eds., *Reconsidering Sputnik: Forty Years Since the Soviet Satellite*, Australia, Canada, France, Germany, India (Harwood Academic Publishers) 2000

Launius, Roger D., and Howard E. McCurdy, eds., *Spaceflight and the Myth of Presidential Leadership*, Urbana, Ill. and Chicago (University of Illinois Press) 1997

Launius, Roger D., and Bertram Ulrich, *NASA and the Exploration of Space: With Works from the NASA Art Collection*, New York (Stewart, Tabori and Chang) 1998

Levine, Arnold, *Managing NASA in the Apollo Era*, NASA Special Publication 4102, Washington, D.C. (NASA) 1982

Loftin, Laurence K., *Quest for Performance: The Evolution of Modern Aircraft*, NASA Special Publication 468, Washington, D.C. (NASA) 1985

McCurdy, Howard E., *Faster, Better, Cheaper: Low-Cost Innovation in the U.S. Space Program*, Baltimore and London (The Johns Hopkins University Press) 2001

McCurdy, Howard E., *Inside NASA: High Technology and Organizational Change in the U.S. Space Program*, Baltimore and London (The Johns Hopkins University Press) 1993

Mudgway, Douglas J., *Uplink-Downlink: A History of the Deep Space Network, 1957–1997*, NASA Special Publication 4227, Washington, D.C. (NASA) 2001

National Aeronautics and Space Administration, *Columbia Accident Investigation Board Report, Volume 1*, Washington, D.C. (Government Printing Office) 2003

Newell, Homer E., *Beyond the Atmosphere: Early Years of Space Science*, NASA Special Publication 4211, Washington, D.C. (NASA) 1980

Portree, David S.F., and Robert C. Trevino, *Walking to Olympus: An EVA Chronology*, NASA Monographs in Aerospace History Number 7, Washington, D.C. (NASA) 1997

Reed, R. Dale, *Wingless Flight: The Lifting Body Story*, NASA Special Publication 4220, Washington, D.C. (NASA) 1997

Roland, Alex, *Model Research: The National Advisory Committee for Aeronautics, 1915–1958*, NASA Special Publication 4103, Washington, D.C. (NASA) 1985

Rotundo, Louis, *Into the Unknown: The X-1 Story*, Washington, D.C. and London (Smithsonian Institution Press) 1994

Schorn, Ronald A., *Planetary Astronomy: From Ancient Times to the Third Millennium*, College Station, Tex. (Texas A&M University Press) 1998

Siddiqi, Asif, *Challenge to Apollo: The Soviet Union and the Space Race, 1945–1974*, NASA Special Publication 4408, Washington, D.C. (NASA) 2000

Siddiqi, Asif, *Deep Space Chronicle: A Chronology of Deep Space and Planetary Probes, 1958–2000*, NASA Monographs on Aerospace History Number 24, Washington, D.C. (NASA) 2002

Swanson, Glen E., ed., *"Before this Decade is Out...": Personal Reflections on the Apollo Program*, NASA Special Publication 4223, Washington, D.C. (NASA) 1999

Swenson, Loyd S., Jr., James M. Grimwood, and Charles C. Alexander, *This New Ocean: A History of Project Mercury*, NASA Special Publication 4201, Washington, D.C. (NASA) 1998

Thompson, Milton O., *At the Edge of Space: The X-15 Flight Program*, Washington, D.C. and London (Smithsonian Institution Press) 1992

Tomayko, James E., *Computers Take Flight: A History of NASA's Pioneering Digital Fly-By-Wire Project*, NASA Special Publication 4224, Washington, D.C. (NASA) 2000

Yeager, Charles, and Leo Janos, *Yeager*, Toronto, New York, London, Sydney, Auckland (Bantam Books) 1985

Further Reading

Many who enjoy history seem to harbor a misleading image of those who write it. For better or worse, historians seem to be perceived as individuals who conduct their work in silence and isolation. While there is certainly some truth to this picture, in reality almost no historical writing is achieved without collaboration. *NASA: The Complete Illustrated History* is no different.

To begin with, I am deeply indebted to Dr. Von Hardesty, a distinguished author in the fields of Russian aeronautics and the cultural history of flight, who had a truly formative influence on this text. His wise and thoughtful suggestions added depth and nuance to a narrative in sore need of both.

I am also very grateful for the eloquent Foreword written for this volume by Dr. Buzz Aldrin. Forever associated with the first human steps outside of our home planet, this combat pilot, test pilot, scientist, astronaut, and international advocate of space exploration is justifiably honored, both by the world community and by his own nation.

At Merrell Publishers, a number of people contributed immensely to this book. First, publisher Hugh Merrell saw the value in a manuscript well outside of his firm's customary publishing interest, but pursued it anyway, and with vigor. The company's Editorial Director, Julian Honer, took the project under his protective care and offered gentle advice, occasional correctives, and outstanding overall guidance. I should also mention three people—Managing Editor Anthea Snow, Art Director Nicola Bailey, and Production Manager Michelle Draycott—who kept this project on schedule and faithful to its initial intent. Editor Sam Wythe is the best kind of collaborator; one who misses nothing, raises questions, and offers imaginative solutions. Hans Dieter Reichert and Paul Spencer of HDR Design produced the beautiful volume that became *NASA: The Complete Illustrated History*. Finally, picture editor Jo Walton presented me with a collection of evocative and meaningful photographs from which to choose. I hope I chose well.

Lastly, I want to acknowledge my wife, Christine Gorn, without whom my name would not be on the title page of this book. She persuaded me to undertake this project when I believed that a host of other commitments made it all but impossible. She proved to me beyond doubt what optimism—and what writing squeezed out on evenings and weekends—can accomplish. I love her for this reason, and many others.

Michael H. Gorn
Simi Valley, California, U.S.A

Acknowledgments

The illustrations in this book have been reproduced courtesy of the following copyright holders:

(b: below; c: centre; l: left; r: right; t: top)

NASA/Ames Research Center: 178, 197

NASA/Dryden Flight Research Center: 43tr, cl, b, 44tr, 45, 46, 47t, 48, 52tl, 54tl, b, 56, 57, 66, 67tr, tc, c, b, 68l, 71br, bl, 84, 85br, t, 85, 104bc, 189, 190-191, 193, 194, 203, 206, 207, 222.

NASA/Glenn Research Center: 52th, 66b, 75r, 161, 241

NASA/Jet Propulsion Laboratory (Caltech): 138-

139, 160, 175, 177t, 240r, 282, 283, 284, 285.

NASA/Johnson Space Center: Front cover, back cover br, 89, 96t, 107, 109, 110, 111tr, br, 122, 123, 125, 127, 128l, c, 133t, 134t, 136b, 137t, 140, 143, 144b, 146br, bl, 149, 150tl, br, 151, 152, 153tl, b, 154l, 155, 200bl, 201, 202, 204, 205, 213b, 216l, 220, 221bl, br, 223, 224, 225t, 226b, 227, 228, 229, 230, 231, 232, 234br, 236-237, 242b, 245b, 246, 247, 248t, c, cl, 249, 251, 255, 260br, bl, 261, 263, 265, 266-267, 268, 274t, 275, 280, 281.

NASA/Johnson Space Center – Earth Science and Image Analysis: 271bl.

NASA/Kennedy Space Center: 94, 95br, 97bl,

102-103, 116, 120, 128r, 130, 135, 142l, 147, 148t, 154r, 159, 162r, 163, 164, 165, 166, 170, 171r, 174, 177, 179r, 180, 181, 182, 183l, 186-187, 209, 211, 212, 213t, 216r, 217, 221t, 233, 235, 238, 240l, 245t, 254, 260t, 269, 274b, 278-279.

NASA/Langley Research Center: 4-5, 12, 14t, 16r, 17, 18, 19tl, 23l, 24bl, 24br, 25t, br, 26bl, 27, 30, 31, 32, 33, 34c, t, br, 37, 38tl, tr, tc, c, cl, 39, 40b, 42, 44tl, 55b, 58, 59t, br, 60, 61, 67tl, 71c, tr, 87, 92b, 114t, 195t.

NASA/Marshall Space Flight Center: 2, 73, 74, 75l, 76-77, 78, 79bl, 83, 88bl, br, 95l, 113tr, 115, 117, 118-119, 121, 142r, 167, 171l, 177br, bl, 184, 188, 195b, 196, 198-199, 200t, br, 208, 218, 242t, 243, 248br, 250, 258,

264, 270, 271t, br, 272, 273.

NASA/HQ GRIN: Back cover tl, tr, cl, cr, bl, 6-7, 10-11, 13, 14b, 15, 16l, 19tr, cl, bl, 20, 21, 22, 23r, 24t, 26r, 28, 29bl, br, 34bl, 35, 36, 38bl, 40t, 41, 43tl, 47b, 49, 50-51, 53, 54tr, c, 55t, 59bl, 62, 63, 64, 65, 69, 70, 72, 75c, 79tl, tr, br, 80, 81, 82, 85bl, 88tl, 90-91, 92c, t, 93, 96bl, 97br, t, 98, 99, 100, 101, 103r, 104tr, cr, br, l, 105, 106, 111tl, 112, 113b, 134b, 136t, 137b, 141, 144t, 146bl, t, 148br, 150, 153tr, 156, 157, 158, 162l, 168, 169, 172, 173, 176, 179l, 183r, 185, 191r, 210, 214, 215, 225b, 226t, 234t, bl, 239, 259, 262, 276, 277, 278l.

NASA/Space Telescope Science Institute: Endpapers, 252, 253, 256, 257.

Picture Credits

Index

Published by Merrell Publishers Limited

Head office
81 Southwark Street
London SE1 0HX

New York office
740 Broadway, Suite 1202
New York, NY 10003

merrellpublishers.com

First published 2005
Paperback edition first published 2008

Catalog records for this book are available from the
Library of Congress.

British Library Cataloguing-in-Publication Data:
Gorn, Michael H.
NASA: the Complete Illustrated History
1.United States. National Aeronautics and Space
Administration 2.Aerospace industries – United States –
History 3.Astronautics – United States – History 4.Outer
space – Exploration – United States – History
I.Title
629.1'0973

ISBN 10 1-85894-427-9
ISBN 13 978-1-85894-427-2

Produced by Merrell Publishers Limited
Designed by HDR Visual Communication
Copy-edited by Christine Davis
Proof-read by Jacquie Meredith
Indexed by Hilary Bird

Printed and bound in China

Michael Gorn is the author of eight books about the
history of aeronautics and spaceflight, including
*Expanding the Envelope: Flight Research at NACA and
NASA* (winner of the Gardner-Lasser Aerospace History
Literature Award). He also received the Alfred V. Verville
Fellowship from the Smithsonian Institution's National Air
and Space Museum. Gorn served as a historian with the
U.S. federal government for 25 years, in the Department
of Defense, the Environmental Protection Agency, and the
National Aeronautics and Space Administration. He is
presently the Ombudsman at the NASA Dryden Flight
Research Center in California.

Front cover: Buzz Aldrin stands in the Sea of Tranquility as
Neil Armstrong freezes the moment with a 70 millimeter
lunar surface camera. This portrait, taken after the two
astronauts became the first humans to walk on the
Moon in 1969, remains one of the most enduring
images in the history of human space travel

Back cover (from top left): as pp. 10–11 below; as
pp. 50–51 below; as pp. 90–91 below; as pp. 138–39
below; as pp. 186–87 below; as pp. 236–37 below

Endpapers: View of the M17 Swan Nebula, taken
by a camera on the Hubble Space Telescope
in April 2002

Page 2: The Space Shuttle *Atlantis* taking flight,
December 2, 1988

Pages 4–5: Turning vanes in the 16 Foot High Speed
Tunnel at Langley

Pages 6–7: Alan Shepard at the Fra Mauro Highlands on
the Moon, *Apollo 14* mission, January 1971 (see p. 148)

Pages 10–11: A P-51 Mustang in the Full Scale Tunnel
(see pp. 40–41)

Pages 50–51: A Bell X-1E being attached to a B-29
mothership (see p. 54)

Pages 90–91: Buzz Aldrin's bootprint on the Moon
(see p. 133)

Pages 138–39: Martian landscape photographed by
Viking 1 (see pp. 172–73)

Pages 186–87: The Orbiter Columbia on the Shuttle
transporter, December 1980, in the run-up to its launch
the following April

Pages 236–37: Kathy Thornton and Thomas Akers
servicing the Hubble Space Telescope (see p. 251)

Pages 292–93: A nebula in the constellation Orion,
photographed from the Hubble Space Telescope in
December 1999